Residential Foundations

Jim Carr

Home Builder Press
National Association of Home Builders
Books to Build On
A Service of NAHB

Home Builder Press®
National Association of Home Builders
1201 15th Street, NW
Washington, DC 20005-2800
www.builderbooks.com

Residential Foundations

ISBN 0-86718-509-0

Cover by David Rhodes

Printed in the United States of America

Library of Congress Cataloging in Publication Data
Carr, Jim K., 1947–
Residential foundations / Jim Carr.
p. cm.
Includes index.
ISBN 0-86718-509-0
1. Foundations. 2. Dwellings. I. Title.

TH2101.C37 1999
690'.937—dc21 99-052183

Quantity Discounts

Quantity discounts for individual Home Builder Press titles are available. Multititle packages are also available for certain books. For further information, please contact:

Home Builder Press®
National Association of Home Builders
1201 15th Street, NW
Washington, DC 20005-2800
(800) 368-5242, ext. 394; fax (202) 822-0391

Check us out online at www.builderbooks.com

11/99 Harlowe/McNaughton 1500

About the Author

Jim Carr is the coordinator of the Construction Program at The Ohio State University where he teaches students the art of residential construction, a position he has held since 1990. Prior to this he was a home builder and remodelor for 17 years. Throughout Mr. Carr's construction-related career he has been active both locally and nationally with the National Association of Home Builders, serving on numerous committees. A strong advocate of builder education, Mr. Carr also has taught numerous classes on residential construction principles and methods.

Acknowledgments

I would like to thank Gene McKown who has motivated me continuously since I started my career in construction. He gave me a start, helped me open my own business, and pushed me to my limits.

The author and Home Builders Press wish to acknowledge the contributions of the following reviewers who assisted with reviews of the proposal or the developing manuscript: Harold Connor, University of Oklahoma; Jay Crandall, NAHB Research Center, Inc.; Mike DeMattel, Southern Illinois University; Michael Eckert, National Ready Mixed Concrete Association; Jamie Farney, Portland Cement Association; Larry Kness; George Muste, National Ready Mixed Concrete Association; Howard J. Rosenberg, GMR Ltd.; Alvin Sacks; Andrea Vrankar, NAHB Research Center, Inc.

Book Preparation

This book was produced under the general direction of Tom Downs, NAHB Executive Vice President, in association with NAHB staff members Bob Brown, Staff Vice President of Knowledge Management; Adrienne Ash, Assistant Staff Vice President, Information Services; Kurt Lindblom, Acquisitions Editor; Barbara Minich, Copy Editor and Proofreader; David Rhodes, Art Director; and Elisa Subin, Production Editor.

Table of Contents

Figure List

Introduction

"Man who improve house before building solid foundation apt to run into very much trouble," Charlie Chan in The Feathered Serpent *(1948).*

A residential foundation is the substructure that distributes the load, or weight, of a home along the earth on which it rests. The design of a residential foundation is dependent on soil conditions, site-specific circumstances, building materials, and local regulations and traditions. Each foundation must be designed for the specific site conditions and the floor plan of the residence.

Today, most residential foundations fall within one of five major classifications: slab foundations, basement foundations, crawl-space foundations, pier-and-grade-beam foundations, and post-and-piling foundations. The following chapters explore the factors that must be considered while planning a foundation and walk through the steps that comprise the construction of each type of residential foundation. A checklist at the end of each chapter summarizes the main points.

Because the foundation of a home rests on the earth, the make-up of the soil at the construction site must be analyzed and understood before the foundation is designed. The unique characteristics of each soil type and how they can affect the stability and longevity of a foundation are examined in Chapter 1, Understanding Soil Conditions. Resources, soil analyses, and radon testing methods are discussed.

Numerous activities must be accomplished before the first shovel of dirt is excavated. From the permit process to locating utilities, from clearing the site to saving trees, Chapter 2, Beginning Excavation, looks at each of these activities with suggestions that will simplify the process. Methods for accurately estimating excavation costs also are included.

The site is cleared, the foundation plan is prepared, and the permits are approved. Now it is time to site the foundation and footings on the building site. Methods that can be used to site and square the foundation are explained in Chapter 3, Siting the Foundation and Footings. Once the foundation is laid out, learn how to mark footings and then how to realign the foundation on the footings after they are in place. Techniques for setting up reference points outside the construction area are presented to help ensure that the foundation remains square while the foundation is placed.

Protecting residential foundations against water damage begins with adequate planning. Ways to identify potential water hazards are discussed in Chapter 4, Planning for Water Control. The difference between damp-proofing and waterproofing a foundation is explained and various products that can be used with residential foundations are presented.

Concrete is a material of choice for builders in many regions of the country. A very versatile building material, concrete is economical, durable, and very strong. Because concrete can be formed into various shapes and finished with a wide variety of textures and colors, builders employ a number of techniques to create basement foundations, crawl spaces, and slabs on grade. Chapter 5 discusses the compressive strength, slump, and curing process for concrete placed in footings and foundations, as well as the finishes that can be used.

Foundation systems meet the needs of builders who must construct a stable home on varying soils according to architectural specifications. Residential foundations can be constructed using a variety of materials, including concrete, concrete blocks, pressure-treated wood, and plastics such as polyethylene foam. Even packed earth has been used as a foundation material. The availability of materials and labor also can influence builders' choices of foundations.

Footings spread the load of the structure over the soil. The specific requirements involved in designing and placing footings beneath residential foundations are explained in Chapter 6, Using Footings in Foundations. The various types of footings and the materials that can be used in their construction are discussed in detail.

Chapter 7, Placing Cast-in-Place Concrete Foundations examines each of the relevant factors that must be considered before pouring a concrete foundation, such as the slump and strength of the concrete. Methods for compensating for exterior pressure with bracing, constructing forms, and curing the concrete also are provided.

Builders have several options to improve the insulating quality of concrete walls. From polystyrene foam sheathing to insulated concrete forms, each of these methods are examined and the method of construction presented in Chapter 8, Building with Insulating Concrete Forms.

While many residential foundations are cast in place at the building site, precast foundation panels are an option. Chapter 9, Using Precast Foundation Panels, answers questions about when these systems are most effective and economical. The equipment needed to hoist the panels and the techniques for tying them together are described in detail.

A basic slab foundation consists of a concrete floor and a footing. When soil conditions require a reinforced slab, monolithic or post-tensioned slab foundations may provide an answer. Both types of slabs are described in detail in Chapter 10, Constructing Slab Foundations.

Concrete masonry units (CMUs), more commonly called concrete blocks, have been used for 100 years to construct residential foundations. The types of blocks, surface finishes, and construction methods are discussed in Chapter 11, Building with Masonry Systems.

When the ground soil on a site is unable to adequately support a home, the foundation system must reach down to stable soils below. Piers, posts, and pile foundations offer alternatives for these situations. The similarities and differences between the three foundation types are explained in Chapter 12, Designing Pier, Post, and Piling Foundations. Construction methods also are described.

Often used to construct foundation forms and temporary bracing, wood also is a material of choice for foundations. First referred to as All Weather Wood Foundations, the advantages of preserved wood foundations are explored in Chapter 13, Building Preserved Wood Foundations. The process and chemicals used to preserve wood, required grade stamps, and construction techniques are presented.

Frost-protected foundations are an alternative for residences built in specific parts of the country under specific conditions. Chapter 14 explores the insulation and processes used to build these foundations.

Chapter 15 summarizes the estimating costs for all the phases involved in placing a residential foundation. From site clearing costs to footings to the actual foundation materials, the information in this chapter helps builders estimate costs by mixing and matching the materials and systems used.

1

Understanding Soil Conditions

When it comes to foundations, not all soils are equal. Excavating is easy in some soils, while other soils constantly collapse or shift once digging begins or when weight is placed on the soil. Generally, to provide adequate support for a typical residence, soil must support at least 1,500 pounds of weight per square foot (psf). As a rule of thumb, if the soil compresses when an adult steps on it, the soil probably does not have an adequate bearing capacity.

Before designing, excavating, or building any foundation system, builders must be familiar with the type of soil at the site, as well as geological and climatic conditions. All three factors influence the design of the foundation and the materials used.

Information on general soil conditions for most U.S. counties is available from the U.S. Department of Agriculture's (USDA's) Soil Conservation Service (SCS). In regions with difficult or unstable soils, a soil survey performed by a qualified geotechnical engineer or foundation engineer is needed to provide highly specific information. The SCS map in Figure 1-1 shows the various soil types in the area. Individual soil types are listed in Figure 1-2 with the associated symbol and some of the related physical characteristics.

As part of the platting requirements for new developments, some municipalities require engineers for land developers to design a typical foundation system for residences that will be built in the proposed subdivision. Engineers may submit or sign off on one design for single-story structures and another for two-story structures. Developers also must indicate which lots have areas that contain fill and the depth of that fill. If the site includes areas of undesirable fill further site exploration is warranted to be sure that an appropriate foundation system is designed.

FIGURE 1-1. The symbols employed on Soil Conservation Service maps, such as the ones shown in Figure 1-2, indicate the various soil conditions found in the county.

[Some terms that describe restrictive soil features are defined in the Glossary. See text for definitions of "slight," "moderate," and "severe." Absence of an entry indicates that the soil was not rated. The information in this table indicates the dominant soil condition; it does not eliminate the need for onsite investigation]

Soil name and map symbol	Shallow excavations	Dwellings without basements	Dwellings with basements	Small commercial buildings	Local roads and streets	Lawns and landscaping
AdD, AdF---------- Alexandria	Severe: slope.	Severe: slope.	Severe: slope.	Severe: slope.	Severe: low strength, slope.	Severe: slope.
BnA, BnB---------- Bennington	Severe: wetness.	Severe: wetness.	Severe: wetness.	Severe: wetness.	Severe: frost action, low strength.	Moderate: wetness.
BrD, BrE, BrF----- Berks	Severe: slope.	Severe: slope.	Severe: slope.	Severe: slope.	Severe: slope.	Severe: slope.
BsB--------------- Bethesda	Moderate: dense layer, large stones.	Slight---------	Slight---------	Moderate: slope.	Moderate: frost action.	Moderate: droughty.
BsF--------------- Bethesda	Severe: slope, slippage.	Severe: slope, slippage.	Severe: slope, slippage.	Severe: slope, slippage.	Severe: slope, slippage.	Severe: slope.
BtA--------------- Bogart	Severe: cutbanks cave, wetness.	Moderate: wetness.	Severe: wetness.	Moderate: wetness.	Severe: frost action.	Slight.
BtB--------------- Bogart	Severe: cutbanks cave, wetness.	Moderate: wetness.	Severe: wetness.	Moderate: wetness, slope.	Severe: frost action.	Slight.
CdA--------------- Canfield	Severe: wetness.	Moderate: wetness.	Severe: wetness.	Moderate: wetness.	Severe: frost action.	Moderate: wetness.
CdB, CdB2--------- Canfield	Severe: wetness.	Moderate: wetness.	Severe: wetness.	Moderate: wetness, slope.	Severe: frost action.	Moderate: wetness.
CdC, CdC2--------- Canfield	Severe: wetness.	Moderate: wetness, slope.	Severe: wetness.	Severe: slope.	Severe: frost action.	Moderate: wetness, slope.
CfB*:						
Canfield---------	Severe: wetness.	Moderate: wetness.	Severe: wetness.	Moderate: wetness, slope.	Severe: frost action.	Moderate: wetness.
Urban land.						
CgB, CgB2--------- Cardington	Severe: wetness.	Moderate: wetness, shrink-swell.	Severe: wetness.	Moderate: wetness, shrink-swell, slope.	Severe: low strength, frost action.	Slight.
CgC, CgC2--------- Cardington	Severe: wetness.	Moderate: wetness, shrink-swell, slope.	Severe: wetness.	Severe: slope.	Severe: low strength, frost action.	Moderate: slope.
Ch, Ck------------ Carlisle	Severe: excess humus, ponding.	Severe: flooding, ponding, low strength.	Severe: flooding, ponding, low strength.	Severe: flooding, ponding, low strength.	Severe: ponding, flooding, frost action.	Severe: ponding, flooding, excess humus.
CnA--------------- Chili	Severe: cutbanks cave.	Slight---------	Slight---------	Slight---------	Moderate: frost action.	Slight.
CnB--------------- Chili	Severe: cutbanks cave.	Slight---------	Slight---------	Moderate: slope.	Moderate: frost action.	Slight.

Source: U.S. Department of Agriculture Soil Conservation Service.

FIGURE 1-2. Soil Conservation Service maps provide information on soil conditions for most U.S. counties.

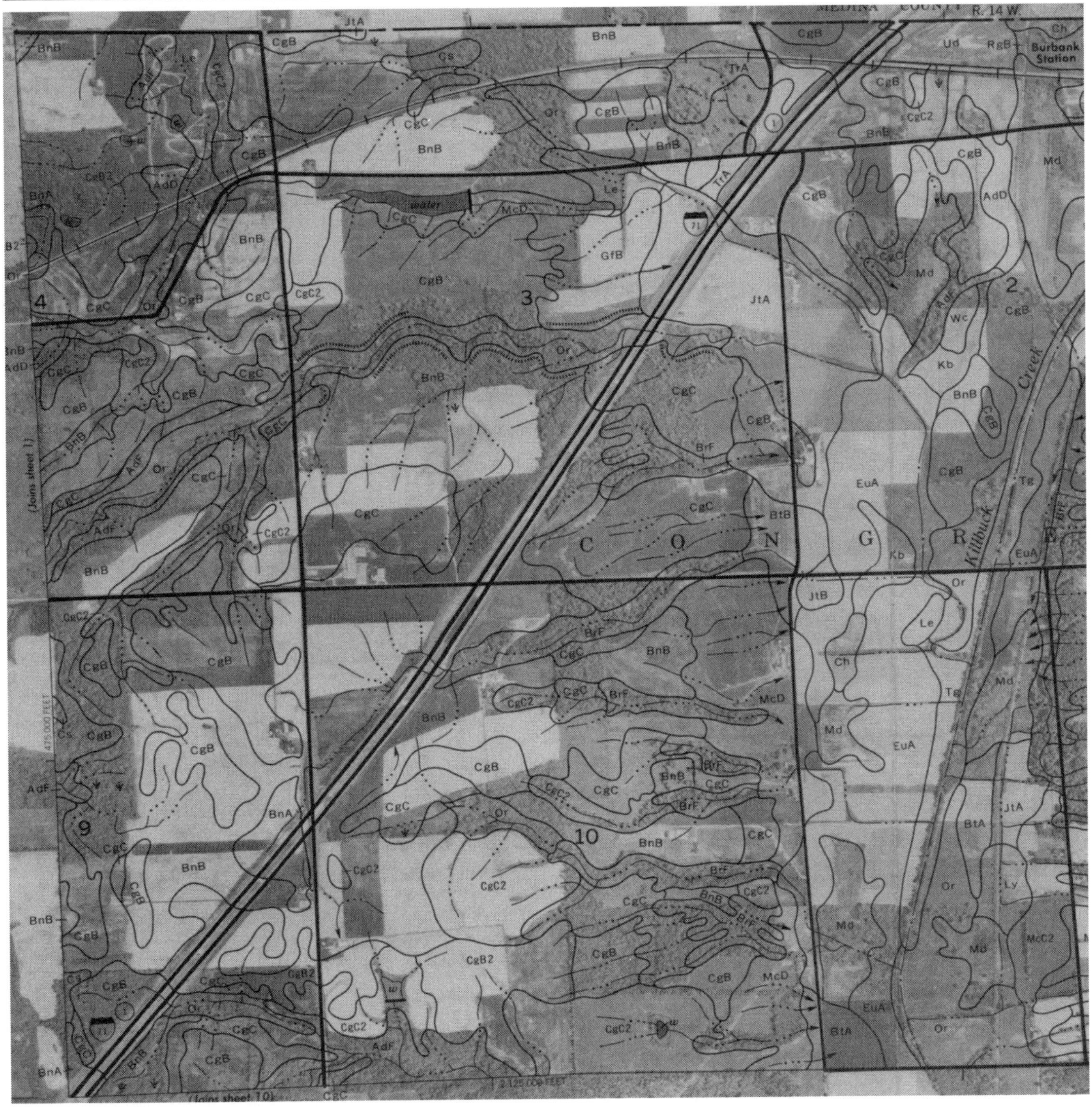

Source: U.S. Department of Agriculture Soil Conservation Service.

Understanding Soil Components

Soil is composed of unconsolidated particles of weathered rock, organic materials, water, and air. As rock weathers, it gradually breaks down into small particles of sand, silt, and clay. In an ideal soil, 40 percent of the volume is comprised of sand, silt, and clay. Air and water account for another 50 percent, and organic materials—mostly decomposed plant material and animal wastes—generally make up the remaining 10 percent. The diagram in Figure 1-3 show soil samples divided into parts.

The differences in the sizes and proportions of the particles of sand, silt, and clay and their resultant chemical interactions create different soil compositions. Variations in the combination of rock particles, organic materials, water, and air also make each soil different. A typical soil profile is shown in Figure 1-4.

Classifying Soils

Soil is classified generally by textural group, and technically by the size and proportion of the rock particles and by how well the soil components hold together. Soils that hold together well are *cohesive* soils, while those that tend to fall apart are called *cohesionless* soils. According to the American Society for Testing and Materials (ASTM), soils can be classified by type and size of particles along a continuum, which is shown in Figure 1-5.

Two of the major soil classification systems used for foundation engineering are the Unified Soil Classification System (USCS) and the American Association of State Highway and Transportation Officials (AASHTO) Classification System. In addition to the percentage of materials and the differing sizes of the rock particles, these soil classification systems use three characteristics, identified through lab testing, to classify soils:

FIGURE 1-3. Soil is made up of rock particles, organic materials, water, and air as shown in this diagram of a soil sample.

Source: Site Engineering for Developers and Builders, *by Thomas L. Brown, Home Builder Press, National Association of Home Builders, Washington, D.C.; 1988.*

FIGURE 1-4. This illustration shows a typical soil profile.

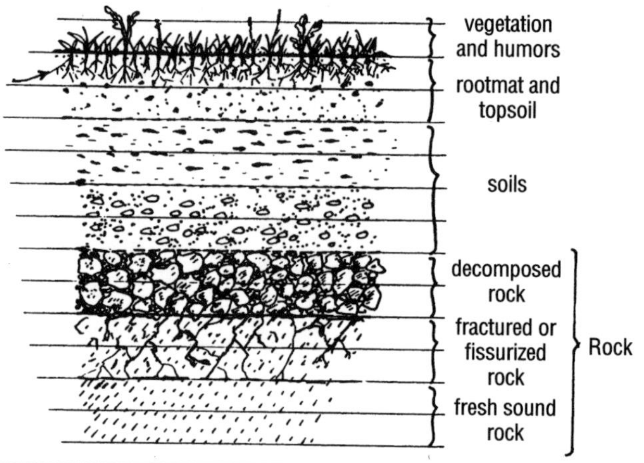

Source: Site Engineering for Developers and Builders, *by Thomas L. Brown, Home Builder Press, National Association of Home Builders, Washington, D.C.; 1988.*

FIGURE 1-5. The American Society for Testing and Materials classifies soils by type and size of particles along a continuum.

Cohesionless	
Gravel	Particles larger than 2 mm
Coarse Sand	Particles between 0.4 and 2 mm
Fine Sand	Particles between 0.8 mm and 0.4 mm
Silt	Particles between 0.005 mm to 0.08 mm
Cohesive	
Clay	Particles smaller than .005 mm

Source: American Society for Testing and Materials.

- *Plasticity* or flexibility is the ability of the soil to be deformed into a different shape.
- *Water-holding capacity* is the ability of the soil to retain water without changing volume.
- *Shear strength* is the soil's ability to withstand certain internal pressures.

Analyzing Soils

Soil analysts look for two main properties, the *load-bearing capacity* or the amount of weight the soil can support, and the soil's susceptibility to *volume changes* or the way in which moisture content and temperature affect the soil. Typical soil types and their general load-bearing capacity are provided in Figure 1-1.

Some soils experience dramatic changes when the moisture content increases or decreases, while others remain almost unchanged. Water flows easily through sandy soils and those with a high gravel content. These soils compact readily and are fairly easy to build on, though they can be susceptible to erosion. Clay soils, however, tend to retain moisture and change as a result. Clay soils with smooth, thin rock particles and low cohesiveness tend to be more elastic. As a result, they may flow when weight is placed on them.

Certain clay soils also change volume when their moisture content changes. As these *expansive* soils absorb water, they swell—and even lift—the structures built on them, exerting pressure that can cause structural stress and damage. Conversely, as these clay soils dry, they shrink and harden, causing the earth to compact around the structure. Because of these moisture-induced volume changes, expansive clay soils are among the most difficult to build on and usually require a specially designed foundation.

Silty soils also can complicate the building process. Dry silty soils are brittle and can fracture under the weight of a structure. Also, because silty soils compress easily the weight of a structure on top of the soil can cause the underlying soil to shift, leading to structural damage.

A USCS chart that can be used to identify soil types of a specific site and indicate the acceptability of that site for construction is provided in Figure 1-6.

FIGURE 1-6. This United Soil Classification System chart identifies soil types by group and acceptability for construction.

SOIL GROUP	UNIFIED SOIL CLASSIFICATION SYSTEM SYMBOL	SOIL DESCRIPTION	DRAINAGE CHARACTERISTICS[1]	FROST HEAVE POTENTIAL	VOLUME CHANGE POTENTIAL EXPANSION
Group I	GW	Well-graded gravels, gravel sand mixtures, little or no fines.	Good	Low	Low
	GP	Poorly graded gravels or gravel sand mixtures, little or no fines.	Good	Low	Low
	SW	Well-graded sands, gravelly sands, little or no fines.	Good	Low	Low
	SP	Poorly graded sands or gravelly sands, little or no fines.	Good	Low	Low
	GM	Silty gravels, gravel-sand-silt mixtures.	Good	Medium	Low
	SM	Silty sand, sand-silt mixtures.	Good	Medium	Low
Group II	GC	Clayey gravels, gravel-sand-clay mixtures.	Medium	Medium	Low
	SC	Clayey sands, sand-clay mixture.	Medium	Medium	Low
	ML	Inorganic silts and very fine sands, rock flour, silty or clayey fine sands or clayey silts with slight plasticity.	Medium	High	Low
	CL	Inorganic clays of low to medium plasticity, gravelly clays, sandy clays, silty clays, lean clays.	Medium	Medium	Medium[2] to Low
Group III	CH	Inorganic clays of high plasticity, fat clays.	Poor	Medium	High[2]
	MH	Inorganic silts, micaceous or diatomaceous fine sandy or silty soils, elastic silts.	Poor	High	High
Group IV	OL	Organic silts and organic silty clays of low plasticity.	Poor	Medium	Medium
	OH	Organic clays of medium to high plasticity, organic silts.	Unsatisfactory	Medium	High
	Pt	Peat and other highly organic soils.	Unsatisfactory	Medium	High

For SI: 1 inch = 25.4 mm.

1 The percolation rate for good drainage is over 4 inches per hour, medium drainage is 2 to 4 inches per hour, and poor is less than 2 inches per hour.

2 Dangerous expansion might occur if these two soil types are dry but subject to future wetting.

Source: Council of American Building Officials (95) Table 405.1.

Soils that retain moisture also are susceptible to volume changes induced by freeze-thaw cycles. When ground moisture freezes and expands, the increased volume of the soil can lift objects resting on it. This condition is known as *heave* or *heaving* and can occur when footings are constructed above the frost line, causing problems to the foundation and ultimately the house itself. The *frost line* is the uniform depth below the surface of the ground at which the moisture in the soil is likely to freeze during a typical winter. Minimum winter temperatures are used to determine the depth of the frost line. Consult local building officials to determine the depth of the frost line on the construction site. To minimize damage from heaving, place footings below the frost line. A map showing weather probability is provided in Figure 1-7.

Compensating for Site Characteristics

Just as soils vary in their composition and characteristics, a single building site may include a mixture of soil types and drainage characteristics. An unstable layer of silt or clay may rest on a stable sand or rock formation. Or more stable soils may rest on top of reactive clays or soils high in organic materials.

Problems associated with mixed soils can be dramatic or subtle. For example, if a layer of wet sand lies beneath more stable surface soils, an earthquake can cause the wet sand to liquefy, intensifying the damaging effects of the earthquake to the structure above. In such areas it is imperative that foundation systems be designed for the seismic conditions.

Another problem with mixed soils is that they provide nonuniform support. Houses built on

FIGURE 1-7. Weather probability, illustrated for the United States, is a factor when determining footing depth.

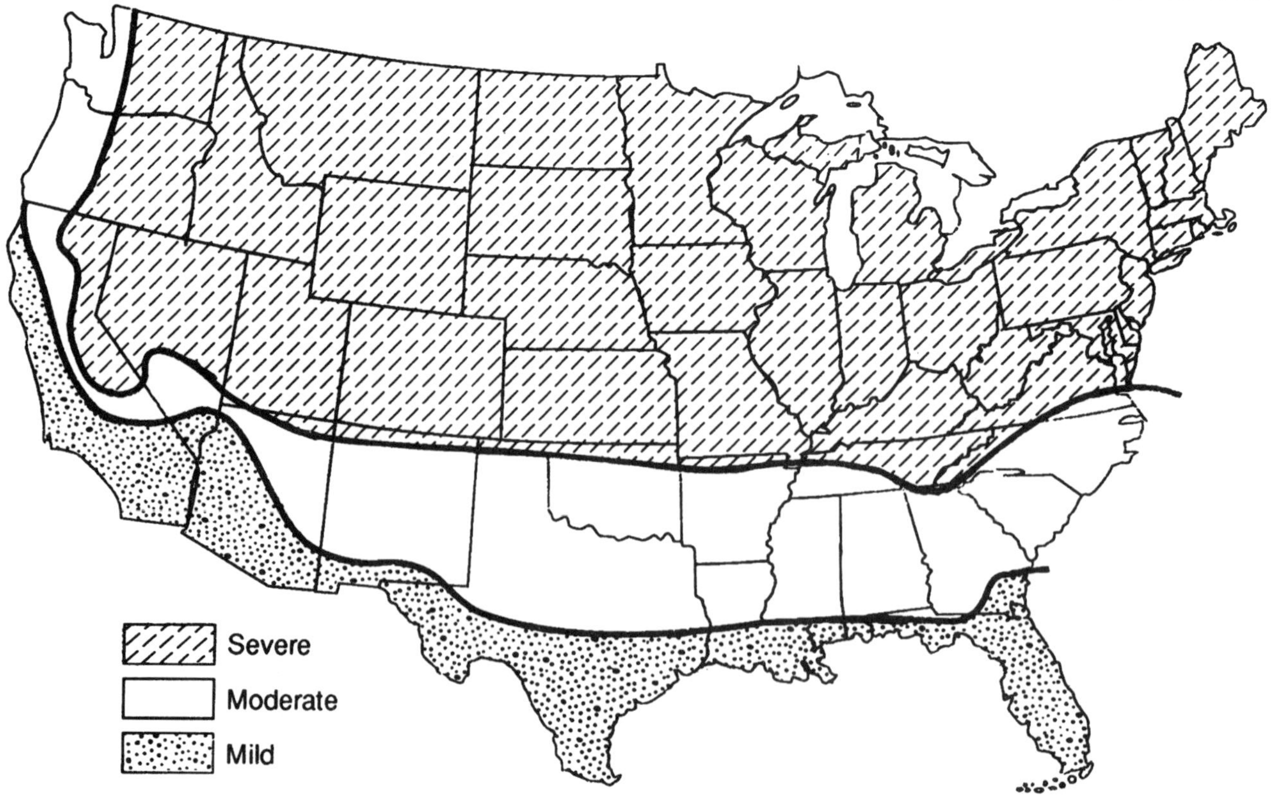

Source: Residential Concrete, *3rd ed., National Association of Home Builders, Washington, D.C.; 1999.*

nonuniform soils may "settle" unevenly, leading to *faulting* or vertical displacement of the slab. Whenever a combination of soil types coexists on a site, the foundation must be designed to meet the site's specific soil and drainage characteristics. The foundation should be designed to meet the requirements created by the least desirable conditions.

Most building codes that dictate footing design are based on average soil conditions for the locality. Unfortunately, soil conditions can vary greatly even within a small area. Even when a structure is built to code, future foundation problems may arise if the dwelling is sited on nonaverage or undesirable soils. It is a good idea to check SCS maps and those provided by the state and municipality to see if the proposed building site is in an area known to have difficult soil conditions. Building officials and other builders familiar with the area also can provide valuable information.

Conducting a Soil Survey

A soil survey is a more exact way to determine specific soil characteristics at a given site. Most building codes state that tests may be required in areas where the soils have the potential to be expansive, compressible, or unstable and thus are classified as hazardous. The testing clause also covers situations when the characteristics of the soil are unknown.

During a typical soil survey a soil or geotechnical engineer will explore the site and take samples to determine the soil conditions. The type and thickness of each layer of soil then is identified and the moisture content and compaction of the soil measured. The survey also can determine the seasonal level of the water table based on the mottled colors in the soil layer. One sampling method involves drilling a series of test holes at the proposed construction site. A second sampling method involves analyzing the sides of a trench excavated specifically for this purpose.

When a detailed subsurface analysis is required, two holes usually provide sufficient information to determine the quality of the soil under the proposed residence. The two holes should be drilled at points opposite each other on the site and just outside the foundation perimeter, unless a specific area is in question. If the investigation is to determine if or how well an area has compacted, one of the holes should be made in compacted soil. Usually the test holes are drilled to a depth of 10 feet for this type of investigation. The cost of the soil survey and the foundation engineering is well spent when compared to the cost of repairing a residential structure caused by an inadequate foundation.

Compensating for Local Features

Rivers and streams on or near a building site create unique problems when locating and designing residential foundations. During the late 1970s the Federal Emergency Management Agency (FEMA) through the National Flood Insurance Program mandated that municipalities regulate construction in flood-prone areas. The U.S. Army Corps of Engineers developed maps for FEMA that delineate the main channels and boundaries of the 50- and 100-year *flood plains* for major streams and rivers in the United States. An example of a FEMA flood map is shown in Figure 1-8.

In most areas, municipal legislation prohibits new construction and significant improvements to structures located in a *floodway* or a zone identified as a flood channel with a fairly high risk for flooding. Local ordinances also may delineate what procedures are permissible in the *flood fringe* or the land located outside of the main channel but inside the 100-year flood plain. Most such legislation requires that all habitable areas of the construction be at least 1 foot above the level of the applicable 100-year flood plain.

Several techniques may be employed to help meet these requirements. For example, importing fill may raise the level of a site. The soil used to provide the fill is placed in *lifts* or layers approximately 6 inches thick, with each layer properly compacted. To ensure the stability of a home built on a heavily filled site, the foundation must incorporate deeper footings or use some other means to transfer the weight of the finished home to properly compacted or undisturbed soil. A pier-and-grade-beam foundation provides one method of accomplishing this.

Another local geologic condition that may affect the selection of a foundation type is the presence of subsurface water, commonly referred to as *groundwater.* When a hole is dug, the level at which the ground is saturated with water is referred to as the *water table.* In most areas, the water table varies by season and the amount of rain, snow, or other precipitation received in the vicinity. A high water table can erode the stability of sandy or silty soils.

FIGURE 1-8. Federal Emergency Management Agency maps delineate the main channels and boundaries of 50- and 100-year flood plains for U.S. counties.

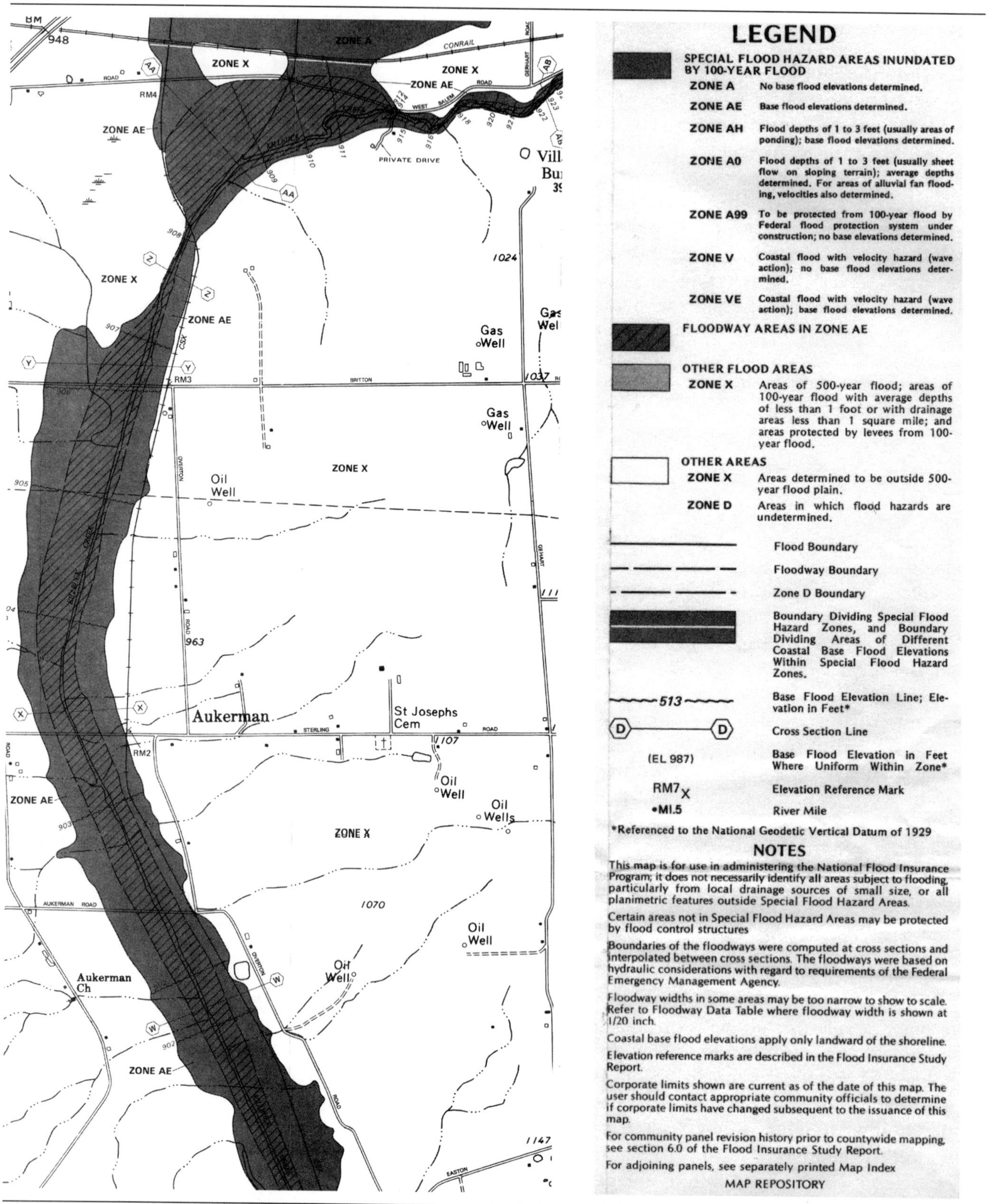

Source: Federal Emergency Management Agency.

When the water table rises, voids in the soil usually filled with air become filled with moisture. The wet earth weighs more, creating an additional load on the surface of the basement wall. This *hydrostatic* pressure can be intensified if the wet soil includes expansive clays. If the water table rises to the level of or above the basement slab, special measures may be required to protect the slab and the foundation walls from water damage and hydrostatic pressures in the soil.

In some areas of the United States a relatively shallow layer of porous soil may lie on top of a non-porous clay layer. This combination of soil types can lead to a *perched* or raised water table. During periods of high precipitation the top layer of soil quickly becomes saturated because the clay layer underneath prevents or drastically slows the drainage process. The excess water can find its way under residential footings and saturate the substrate under a slab or collect on the surface of the ground beneath a crawl space. This moisture also can migrate into the house, increasing the humidity, encouraging the growth of mildew, and increasing the moisture content of the lumber to levels that promote decay. Incorporating drains into the footing design can alleviate problems associated with a perched water table, illustrated in Figure 1-9. Additional information on footing drains is provided in Chapter 4.

Encountering Rock

Even after a survey and visual inspection of the construction site, a builder may encounter rock that must be blasted before the foundation can be excavated. If the area is rocky or rock is suspected, and the home will have a basement, be sure to include a *rock clause* in the contract. A typical rock clause reads like this:

> Any and all additional expenses incurred prior to, during, or after excavation due to any construction site condition including, but not limited to, rock, sand, stone, muck, gravel, filled materials, water—underground or otherwise, shifting sand, unstable materials, or any other unusual conditions shall be at the owners' sole expense and *not* part of this contract. Additional charges shall be based on actual cost of this additional work plus (x) percent.
>
> —Courtesy *Wayne Homes.*

Including such a clause in the contract can help provide economic protection if unexpected rock is discovered when excavating for the foundation. Before adding such a clause to a contract, however, ask an attorney to evaluate the wording to be sure it is consistent with local laws and regulations.

FIGURE 1-9. As illustrated, a perched water table forms when the top layer of soil becomes saturated and the clay layer below prevents drainage.

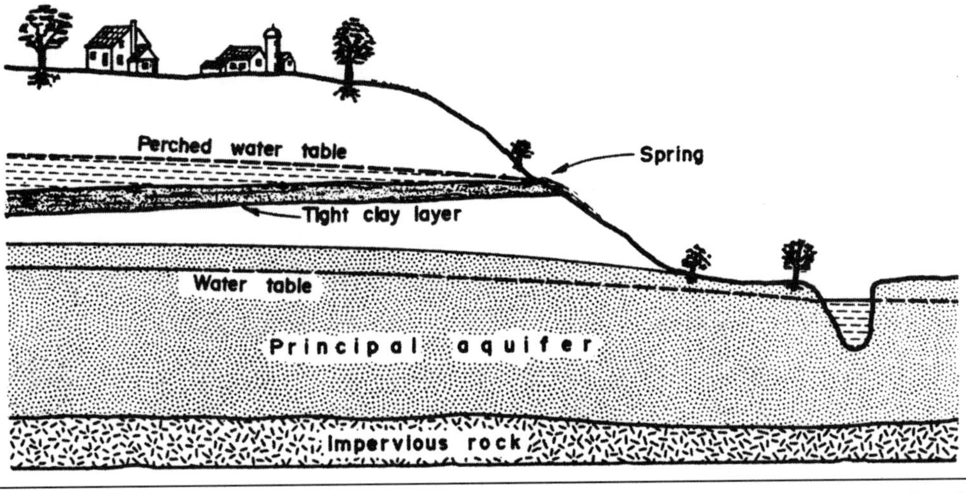

Source: Ground Water and Wells, *ed. by G. F. Briggs and F. G. Fiedler, Wheelabrator Engineered Systems—Johnson Screens, Inc., St. Paul, MN; 1975.*

Testing for Radon Gas

Soils in many parts of the United States overlay naturally radioactive bedrock like granite, black shale, or phosphate-bearing ores that may contain *radium.* A by-product of the decay of radium is *radon gas.* If a home is in a region susceptible to this phenomenon, it is necessary to control the infiltration of this gas into the residence. A map showing regions susceptible to radon gas is shown in Figure 1-10.

Fortunately, most methods for dealing with potential radon gas infiltration are straightforward. When building basement foundations, seal all joints and cracks between the basement floor and the foundation wall with caulk. This includes the joint between the basement slab and the basement wall and any openings, such as chases for pipes, or penetrations through the wall and floor. For slab-on-grade foundations, use similar procedures to seal all protrusions through the slab. For crawl spaces, seal all protrusions through the floor and provide gaskets to seal all access doors. If the crawl space is mostly or completely enclosed, place vents in the foundation walls. Placement of a concrete cap on the soil beneath the crawl space also should be considered. Figure 1-11 is a drawing of radon gas infiltration points in a foundation.

Another option for solving a radon gas problem, which can be used in all foundation systems, is to install mechanical systems that exhaust all gases to the outside. Typically, collection piping is installed under the basement slab and connected to an exhaust vent that carries the gas away through the roof. If testing before or after a structure is completed discovers adverse levels of radon gas, an exhaust fan can be added to the system to enhance its effectiveness and force the gas out of the residence. Figure 1-12 is a drawing of a radon gas removal system.

FIGURE 1-10. Radon gas susceptibility throughout the United States is illustrated in this map from the U.S. Geological Survey.

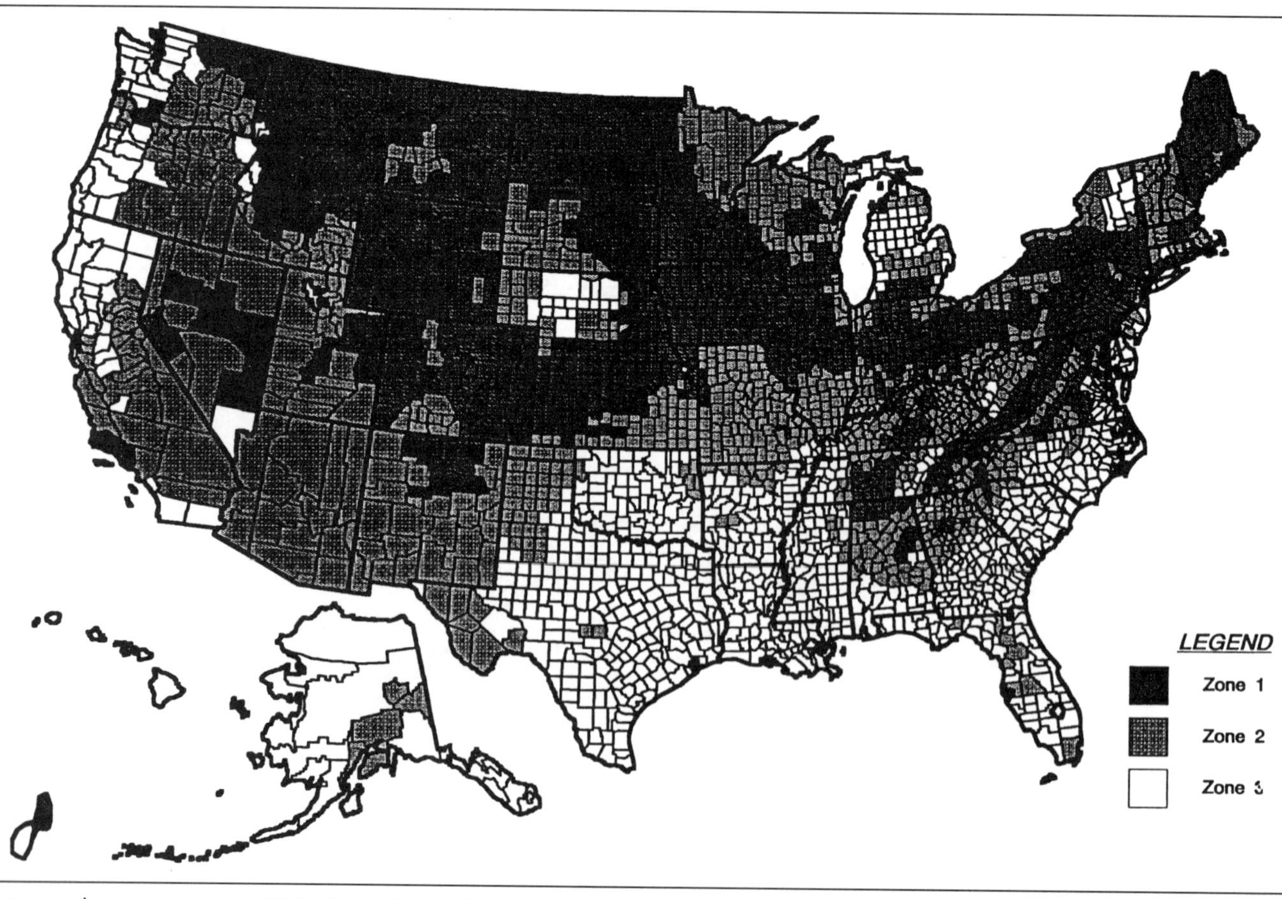

Source: sedwww.cr.usgs.gov:8080/radon/usinpot.gif.

FIGURE 1-11. This illustration shows radon gas infiltration points in a foundation.

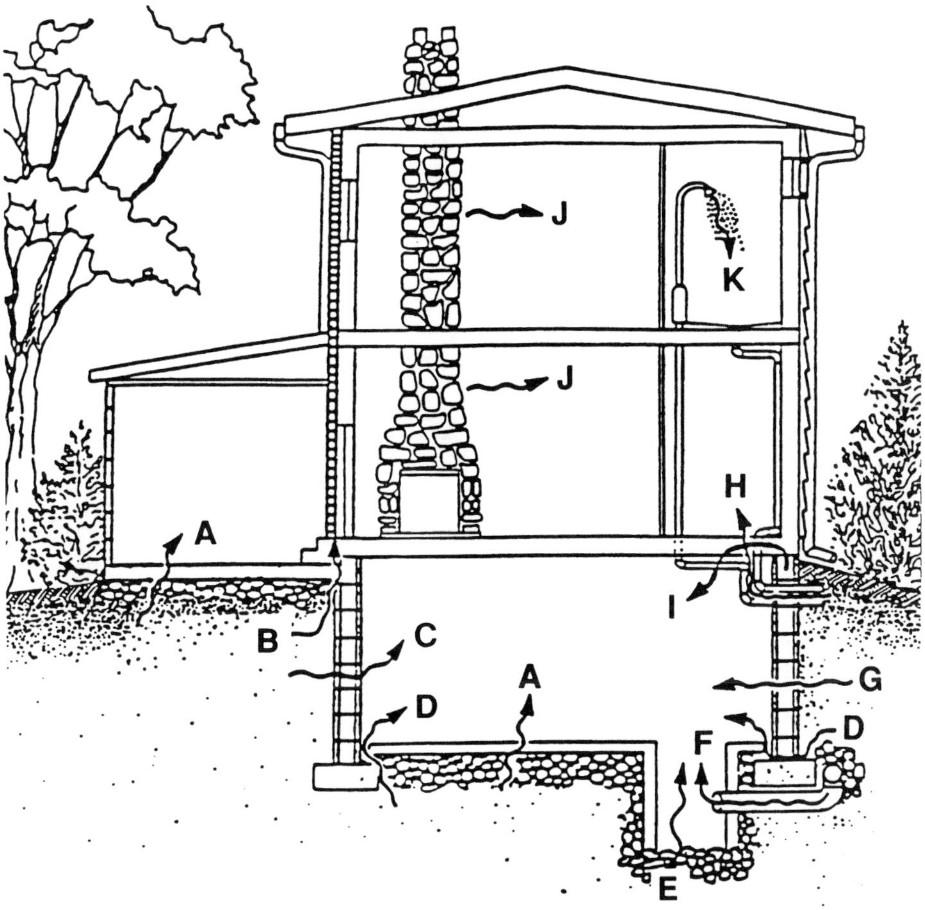

A. Cracks in concrete slabs
B. Spaces behind brick veneer walls that rest on an uncapped hollow-block foundation
C. Pores and cracks in concrete blocks
D. Floor-wall joints
E. Exposed soil, as in a sump pump
F. Weeping (drain) tile, if drained to an open sump pump
G. Mortar joints
H. Loose-fitting pipe penetrations
I. Open tops of block walls
J. Building materials such as some rock
K. Water (from some wells)

Source: Environmental Protection Agency, Radon Reduction Techniques for Detached Houses, Technical Guidance, *2nd ed., EP-88, Washington, D.C., 1988.*

It is extremely difficult to eliminate the infiltration of all gases. Designing and building a foundation using preventive or mitigating measures, however, helps minimize the accumulation of the gas in the house and keeps the levels of those gases that do accumulate within the safe range.

The specific components that affect indoor air quality can be detected and measured using highly sensitive, specialized devices. Radon gas levels can be measured using a *gamma radiation detector.* An *alpha track detector* contains a radiation-sensitive plastic that can be "read" under a microscope to determine concentrations of radon gas. An *electric*

FIGURE 1-12. Sealing radon gas infiltration points or installing a gas removal system, such as the one illustrated, can lower the levels of radon gas in a residence.

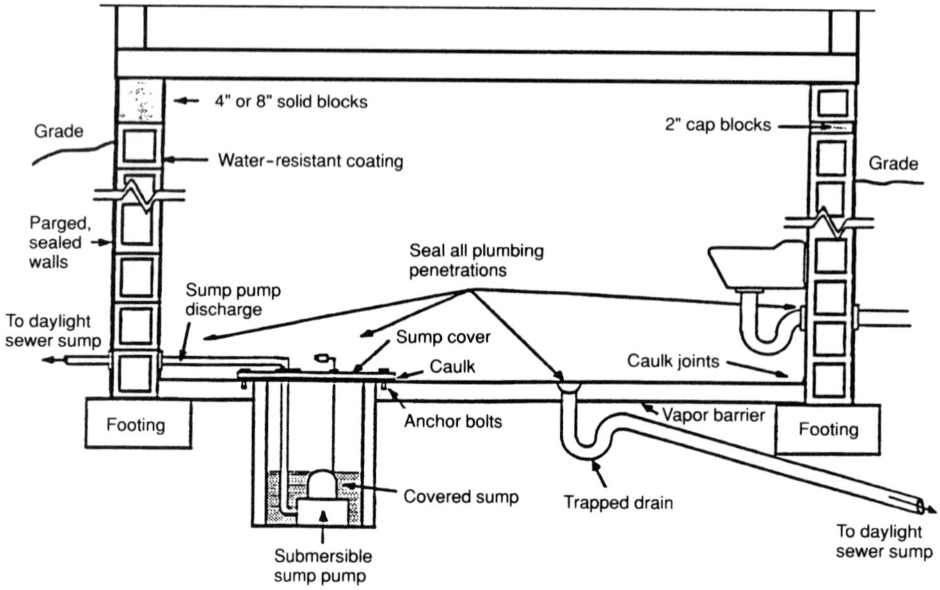

Source: Environmental Protection Agency, Radon Reduction Techniques for Detached Houses, Technical Guidance, *2nd ed., EP-88, Washington, D.C., 1988.*

ion chamber contains an electronically charged disk that when exposed to radon gas reduces its charge. A specialized *voltmeter* then is used to determine the drop of the charge and calculate the concentration of radon gas. *Charcoal monitors* absorb radiation. The amount absorbed then is measured to calculate the concentration of radon gas present.

Because levels or concentrations of radon or other soil gases can vary from season to season, the most accurate assessments of indoor air quality require that the tests be repeated over a period of time. Also, it can take time for concentrations of such gases to build up to detectable levels; therefore, a test performed on a brand-new home immediately after construction is completed may yield a different result from a test on the same home a year or more later.

Chapter Checklist

When assessing foundation types relative to soil, geologic, climatic, and other conditions, consider the following questions.

- Is the lot properly zoned?
- Will the bearing capacity of the soil or the presence of unstable soils require a special footing design?
- Does the building site contain soil that is not compacted?
- Is the area susceptible to potentially hazardous geologic or climatic conditions such as earthquakes, landslides, flooding, or severe weather such as hurricanes or tornadoes?
- How high is the water table at the site?
- Is infiltration of soil gases such as radon gas a consideration for this site?
- What traditional or innovative foundation types have been used in this locality?
- Does this locality require special permits or impose regulations that affect the foundation type?
- Is the site in an area that has the potential to flood?
- Does the prospective buyer prefer one type of foundation to another for aesthetic reasons?

2

Beginning Excavation

Many factors combine to determine the way a house will be situated on a building lot. Zoning ordinances, building codes, easements, deed restrictions, and yard setbacks define specific parameters. The location of utility lines, required well and septic systems, and other factors such as solar orientation also affect the way a house will be sited. Water flow is another consideration. The house should be positioned so that water flows away from the residence without discharging onto adjacent lots and creating problems for the neighbors. In addition, local building traditions and permit processes contribute to the foundation design and placement choices.

Preparing for the Permit Process

Most applications for building permits are accompanied by fairly detailed plans of the proposed structure and usually include a plot plan or survey plat, elevations, foundation plan, floor plan, and section drawings. The plot plan includes a legal description of the property along with a drawing that identifies the boundaries of the building site, the placement of all building lines, rear- and sideyard setbacks, any easements, the footprint of the house, and any walkways or driveways. Figure 2-1 is a typical plot plan.

In areas that have wells and septic systems the local health department may require a copy of the plot plan. This copy of the plan must include the location of the well and septic system on a lot map that also shows the contour of the land. In addition, a copy of the perculation test usually is required. The residence should be at least 5 feet from the septic tank and the absorption bed. The well must be placed 50 feet from the septic tank and 100 feet from the absorption field.

Take the time to double check the plot plan before submitting the plans with the permit application. Making sure that the plan conforms to

FIGURE 2-1. The plot plan, part of the detailed plans for the proposed structure, provides a legal description of the property.

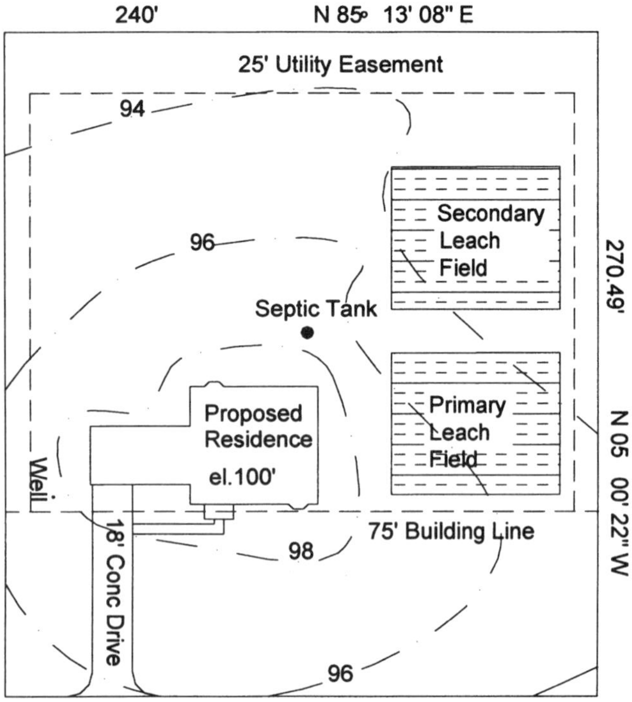

all requirements for the site will save money in the long run. A checklist for reviewing the plot plan is included in Figure 2-2.

Drawing the Foundation Plan

The foundation plan describes the foundation visually by showing the placement of the structural parts and in writing by providing a description of each item. If cast concrete piers are used, for example, the diameter, depth, steel reinforcement, and strength of the concrete are indicated on the plan. Likewise, the width, depth, and placement of reinforcing steel are given if concrete footings are used. The foundation plan for a preserved wood foundation describes the foundation depth, width, and gravel to be used. Parts of the structure, such as pilings, piers, or footings that are underground and not visible are indicated by a series of dashed lines.

Information concerning what will be done to the soil under the residence is detailed on the foundation plan. The type and depth of fill is indicated along with the thickness of any concrete that will be placed. Any reinforcing wire, vapor barrier, or soil treatment to prevent termite or other insect infiltration is indicated and specific information provided. If a preserved wood floor will be used, the size and spacing of the joists and the type of flooring are indicated. The foundation walls also are described, including the thickness and height of the walls and the construction materials. The placement of both vertical and horizontal reinforcing steel bars, if required, is shown.

Along with the structural information about the foundation the placement of plumbing usually is indicated. In residences with a basement, information regarding the structural framing of the dwelling also is provided on the foundation plan, including the direction, size, and spacing of the floor joists and the size and placement of beams and columns. A foundation plan is illustrated in Figure 2-3.

In addition to the actual foundation plan, a *section drawing* usually is required. The section drawing may present a detail of the foundation alone or be part of the section for the complete residence. This drawing shows the vertical placement of the foundation elements. Reinforcing steel is indicated in concrete footings and block walls. For preserved wood foundations the insulation and exterior and interior finishes are illustrated. The placement of the drain pipe and sump pit, and the waterproofing method also are described. Figure 2-4 is a foundation section drawing. Figure 2-5 is a checklist for a foundation plan and section drawing.

FIGURE 2-2. Making sure the proper items are on the plot plan can ease the permit process.

Are the following items included and correct for the zoning and proposed use of the building site?

- Legal description
- Lot dimensions
- Bearings of lot lines
- Building setback lines
- Sideyards
- Easements
- Are the square footage requirements within zoning guidelines?
- Are the driveway and sidewalks indicated?
- Is there sufficient off-street parking?
- Does the size of the backyard meet zoning requirements?
- Are the well, septic tank, and leach field shown, if needed?

FIGURE 2-3. A foundation plan includes information regarding the structural framing of the dwelling.

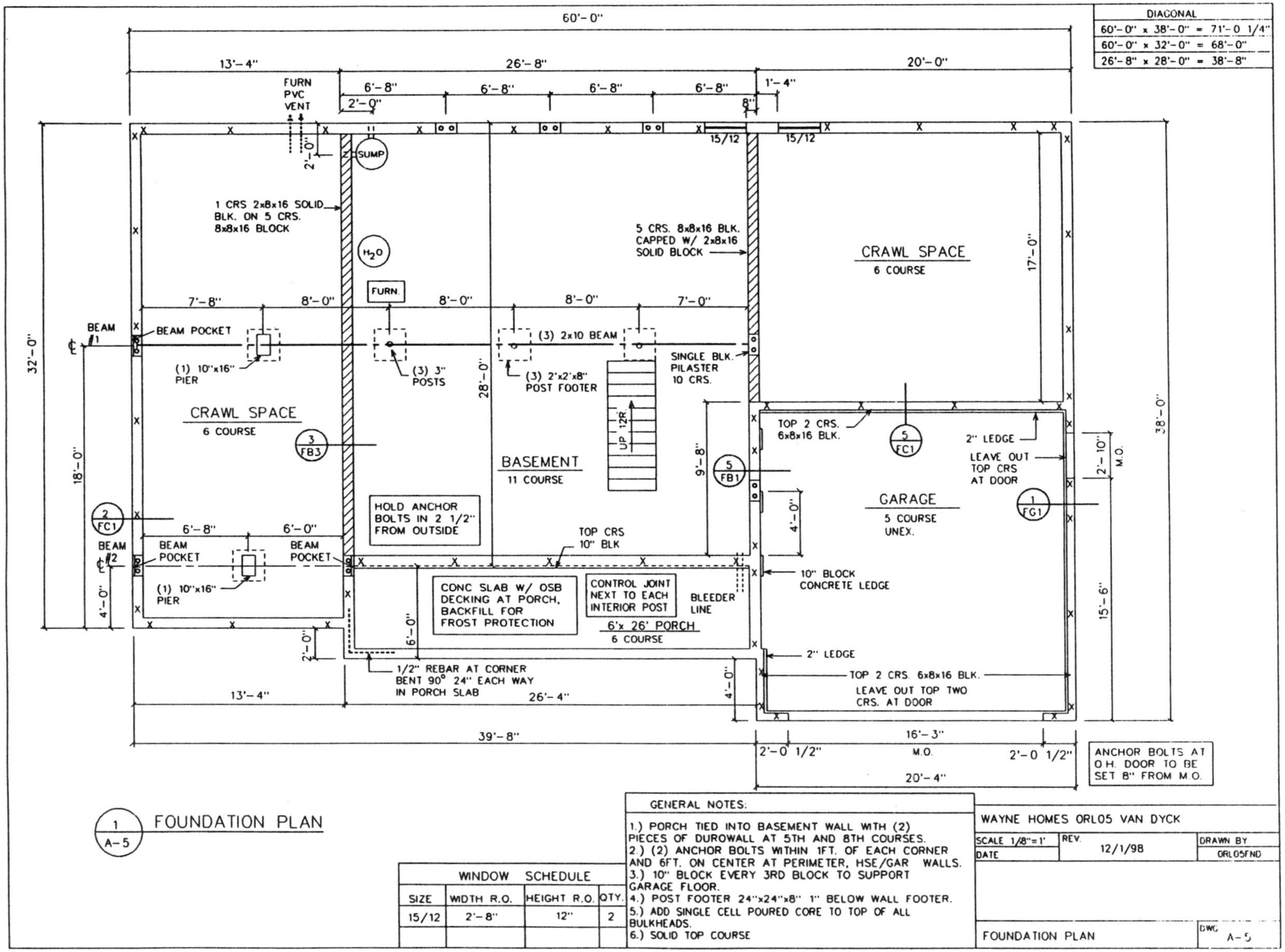

Source: © Wayne Homes, Uniontown, OH.

FIGURE 2-4. The section drawing presents details of the foundation, including the placement of foundation elements.

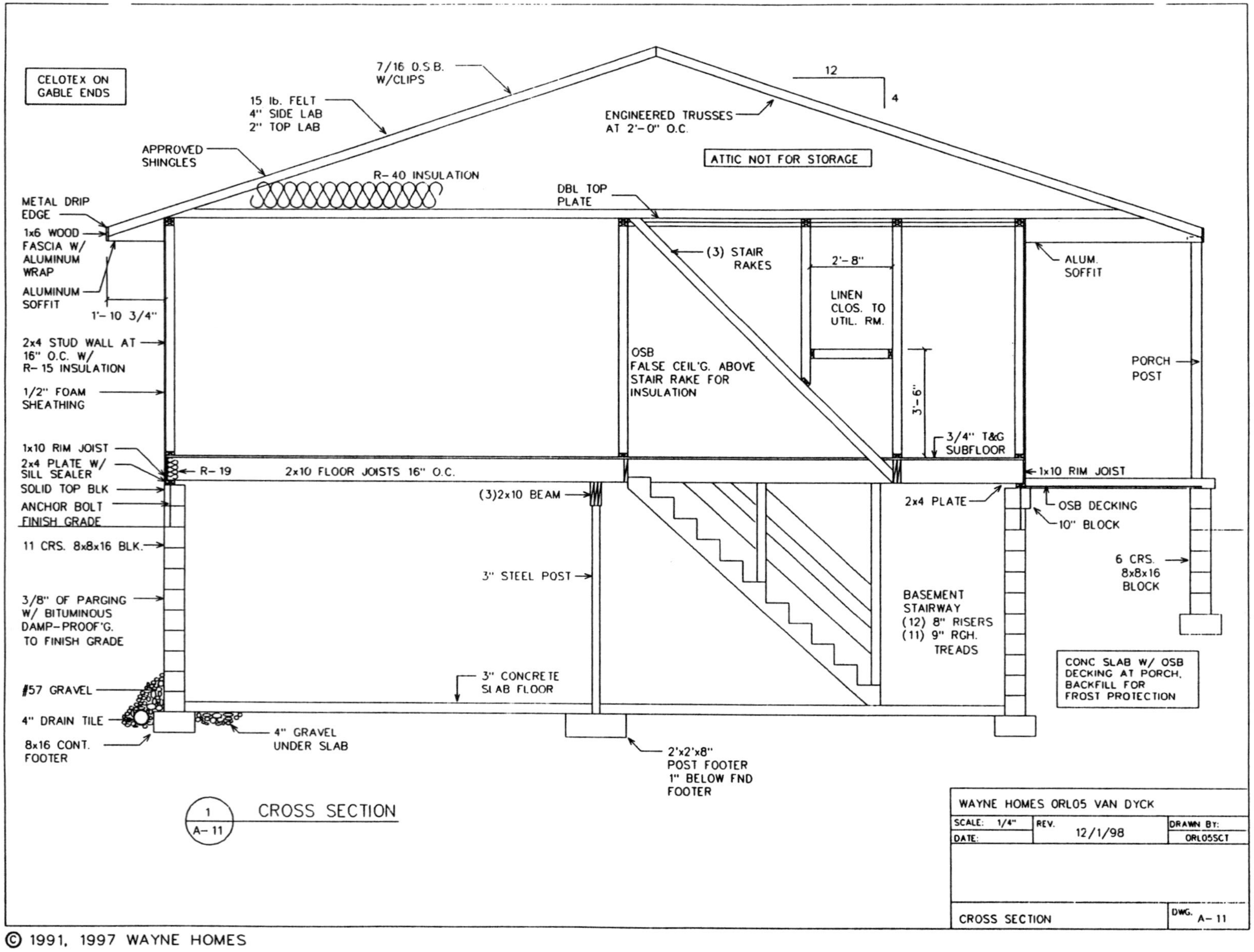

Source: © Wayne Homes, Uniontown, OH.

FIGURE 2-5. Verifying that the foundation elements in the section drawing meet codes before building can save time and money.

Does the foundation plan contain the following information?

- Footings correctly sized for the site conditions
- Foundation walls sized for the site conditions
- Dimensions for all walls
- Diagonal dimensions to help when squaring the foundation
- Spot footings correctly spaced for the described girders
- Correctly placed plumbing fixtures

Clearing the Site

The very first step toward siting a home on a lot is to verify that it is the correct lot. While this may seem an obvious task, pinpointing the boundaries of a building lot is not always easy. It can be very embarrassing and expensive to start building in the wrong spot.

Typically, wood or steel pins or iron pipes are driven into the ground to indicate the corners of individual lots in subdivisions under development. A taller wooden stake with a ribbon attached usually is placed near the corner pin, shown in Figure 2-6, to help locate the individual pins. If the lot corners are not staked, ask the developer to locate the corners. A registered land surveyor should stake out the lot corners of all infill or rural lots. Don't count on a fence to be a true indicator of property lines. There are no guarantees that it was erected correctly.

Connect each of the corner stakes with inexpensive twine or nylon line. The string or line provides a tangible, visible definition of the lot lines, reducing the likelihood that any easements or setbacks will be encroached on during construction.

Protecting Designated Trees

The general rule of thumb is that all existing surface vegetation within 20 feet of the foundation is cleared away before the precise footprint of the residence is laid out or excavation started. Customers, however, may request that specific trees be saved on their lot. Saving selected mature trees can enhance the aesthetic appeal of a site and improve the energy efficiency of the home by providing summer shade and wind breaks in the winter. Many customers consider the presence of mature trees on a site a sign that they are dealing with an environmentally sensitive builder.

If the trees that are to be saved stand too close to the planned foundation, experts can be hired to move the trees to a safer location. Moving trees entails expense and some risk, however, as the trees will experience trauma during the process. Discuss the costs and risks with the customer and ask him or her to make a decision about whether to move the existing trees or replace them after construction with good quality fast-growing trees suitable to the site.

If the owner requests that the trees be left in place, the builder should obtain a written statement from the owner indemnifying him or her from any damage that might be caused to the foundation because of the tree roots. Then identify all trees that are to be saved by placing a price tag on each tree. This will serve as a visual reminder to the workers of their value. Figure 2-7 is a photo of a tree with a protection device.

Take special precautions with the trees that are to be saved to increase the likelihood that they will

FIGURE 2-6. A wooden stake located near a corner pin provides a reference point for finding the other pins that mark the property borders.

FIGURE 2-7. Place a price tag on and a protective device around each tree that the owners want to preserve.

survive construction. Trees can be damaged severely and even die from disturbances to their root system such as those caused by excavation or from ground compression by heavy equipment moving on or off the building site. Trees also can be damaged if a layer of dirt is spread over the ground above the root system or piled up around the tree's trunk, essentially choking it.

The root system of most trees extends out about as far as the branches do. Generally, the more extensive and closer to the tree trunk the disturbance to the root system, the more likely the tree will suffer permanent damage. Young, vigorously growing trees can withstand damage to their root systems better than mature trees.

It also is important to keep equipment away from each tree's root zone. The roots of most trees are 6 to 8 inches below the surface. To provide effective protection, create an equipment-free zone around each tree. Measure the diameter of the tree trunk. Then extend the root protection zone 1 foot out from the trunk for every inch of diameter. For example, keep all construction equipment and vehicles 12 feet away from a tree trunk that is 12 inches in diameter.

More information about trees and their relationship to residential construction is available in *Building Greener Neighborhoods: Trees as Part of the Plan* published by American Forests and the National Association of Home Builders (NAHB); Washington, D.C.; 1995.

Removing Vegetation

While some trees may be left inside the 20-foot clear zone if they will not interfere with the structure, removing trees and other surface vegetation in the vicinity of the foundation helps prevent numerous potential problems from occurring.

An easy way to remove trees and their roots is to push them over with an excavator and then cut them up. If the excavator can apply leverage high on the trunk, often the tree will topple and pull the root system out of the ground. Even trees such as oaks with trunks greater than 24 inches in diameter can be toppled if some of the earth around the root system is removed and the excavator is able to push on the trunk. After the tree is felled, cut it up for removal. The smallest branches can be chipped into mulch with a chipper.

When removing trees, be sure to include the stumps. Stumps left in place decay, creating channels and voids that allow water to penetrate deep into the soil. Eventually the surface soils around the decayed stump will collapse to fill the voids. If the collapse occurs near or underneath a home's foundation, subsidence and structural problems can occur. In the yard, the same process can result in unsightly and potentially dangerous depressions in the ground.

In addition to trees and stumps, remove vegetation such as shrubs, grass, and other plant life. The upper layers of soil that support active vegetation and contain significant amounts of organic materials, otherwise known as topsoil, also should be removed. If the vegetation contained in topsoil is left in place under a foundation it will die and decompose, possibly creating voids under the foundation and undermining its stability.

If at all possible, save the topsoil for use in the final grading and landscaping by stockpiling it on the site. Locate topsoil stored in a "spoils pile" at least as far away from the edge of the excavation as the depth of the excavation to prevent the weight of the topsoil from causing the excavation walls to collapse. Place all dirt piles in a location that will not obstruct construction traffic, interfere with the storage of materials, or cause excessive silt run off.

Remove from the site all *subsurface earth* or *subsoil* not needed for fill as soon as possible to keep the poorer quality soils from mixing with the fill. If not properly compacted, these mixed soils will settle, resulting in depressions around the foundation that might require numerous expensive callbacks to fill.

Removing these mounds of dirt also is a safety consideration. Large mounds of earth can be attractive, especially to children. If children can gain easy access to the building site, cover the topsoil pile with a tarp and take other measures to protect the stockpiles. Discuss with an attorney the appropriate signs and methods for closing off the construction site to protect possessions and limit liability.

Controlling Erosion

Excavated soil is extremely vulnerable to erosion. During the normal construction process one to two truckloads of soil can erode from the jobsite. Take prompt measures to protect the excavation and the surrounding area after topsoil and vegetation are removed.

Prepare a drainage plan for each construction phase. Immediately after the site is cleared of topsoil and vegetation, make provisions for runoff. Provide an avenue for water to move away from the site. Dig *drainage swales*, low channels that direct the flow of excess water, when the lot is scraped to direct the water flow away from the house during construction. The entire lot does not need to be scraped to cut in the swales. The goal is to provide a way for water to flow away from the house with minimum erosion and sediment runoff. This reduces mud buildup and keeps construction materials and the jobsite cleaner, all of which reduces costs and increases safety.

Excavations for basements create a potential for greater problems. A pump may be needed to extract water from the excavation at least while the foundation is being constructed. Slope the pad to the corner of the basement where the sump pump will be located. The slope also will be helpful when installing the underslab drainage system for the house.

If permanent landscaping will not be installed for a long period of time, apply a hydro-mulch with annual grass to decrease potential erosion and keep workers and materials out of the mud. Fabric mesh silt fences, properly installed in trenches dug 4 inches into the soil and placed along contour lines, can retard the action of sheet erosion. Figure 2-8 shows a properly installed silt fence. In addition, construct a storm sewer inlet protection system to prevent sediments from migrating into the storm water system. Figure 2-9 shows an inlet protection system.

FIGURE 2-8. Silt fences made of fabric mesh help retard the erosion of soil from the property.

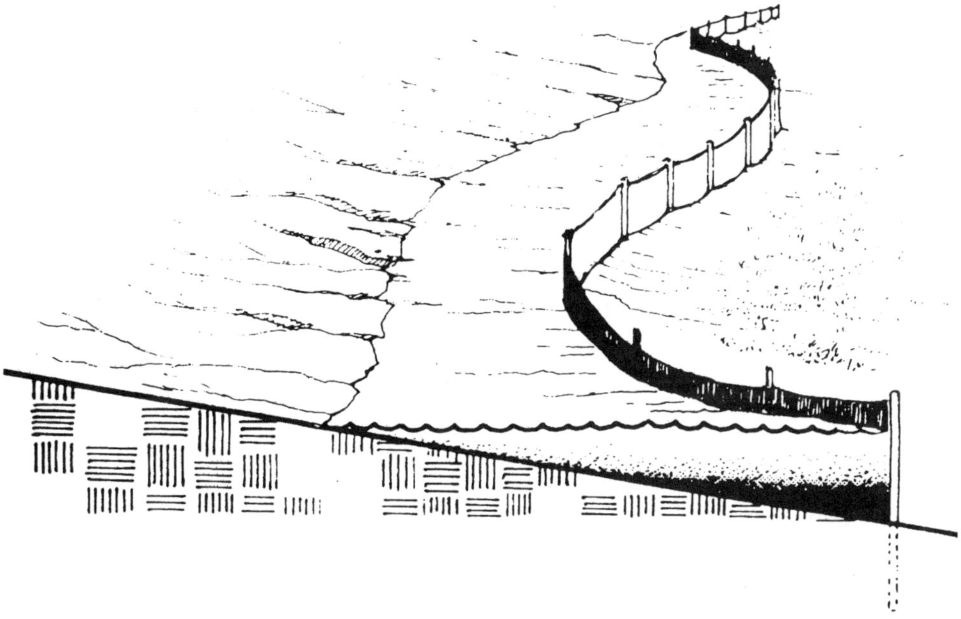

Source: Keeping Soil on Construction Sites: Best Management Practices, *Ohio Home Builders Association and Ohio Department of Natural Resources, Columbus, OH; 1991.*

FIGURE 2-9. An inlet protection system can prevent sediments from entering the storm water drainage system.

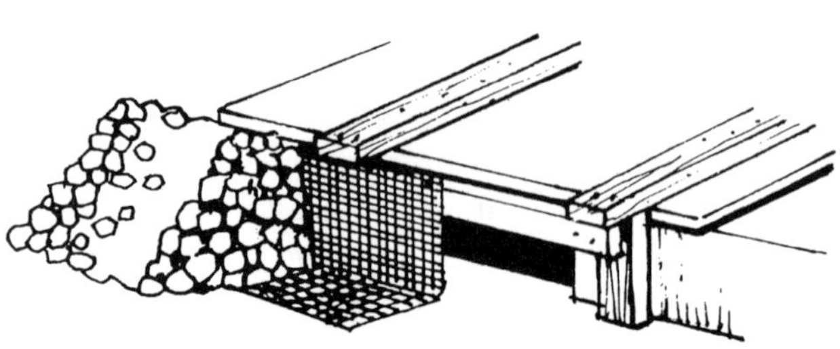

Source: Keeping Soil in Construction Sites: Best Management Practices, *Ohio Home Builders Association and Ohio Department of Natural Resources, Columbus, OH; 1991.*

Managing Construction Waste

In many areas of the United States, builders are encountering escalating landfill costs or surcharges and increasing restrictions on the disposal of construction wastes. Burning, for example, should not be attempted before the appropriate permits are obtained from local municipal authorities. Fortunately, much of the waste materials generated in excavation and foundation preparation can be conserved, reused, or recycled in one way or another. Some creative builders even have found ways to recoup part or all of the costs of disposing of site wastes. For example:

- Lumber mills may purchase the standing timber on the building site.
- Cut smaller trees into firewood and sell or offer it as an amenity to the home buyer.
- Chip smaller tree branches, shrubs, and other vegetation to create landscape mulch. Use the same chipper to dispose of scrap from untreated framing materials. Use the mulch at the jobsite, sell it, or give it to other homeowners in the area.
- Use the spoils from excavations to fill low-lying areas on the site. Note, however, that the filling of potential wetlands may require a permit from the U.S. Army Corps of Engineers or the Environmental Protection Agency (EPA).

Locating Underground Utilities

At least 72 hours before beginning any excavation contact the proper local organization to verify the location of underground utilities near or on the jobsite. In many states, "one-call" organizations act as clearinghouses for the various underground utilities. These organizations, which go by different names, accept responsibility for spotting and marking the underground utilities at the jobsite. Most of these organizations require a proper legal description of the site, such as that included on the plot plan. Omission of this simple step can cause significant and costly delays, and also can be quite dangerous if the excavation proceeds and workers break into buried electrical or gas lines.

The one-call organization will need the following information about the proposed project:

- County
- City and/or township
- Street address
- Intersecting streets or roads and an approximate distance and direction from the intersection to the work
- Extent and location of the work, e.g., frontyard, both sideyards
- Date of work
- Start time
- Type of work
- Company name
- Contact person
- Contact phone number
- Any additional information that might be relevant.

Most one-call organizations record the phone call and give the caller a receipt number for the contact. Record this number and refer to it if the organization must be contacted again, such as when there are changes to the project. Referring to the receipt number shows that the one-call organization was contacted earlier and verifies that the organization was called to check the site.

Mark the proposed site of all excavations with white flags or paint before notifying the one-call organization. Workers from the utility companies usually will mark the entire site, as shown in Figure 2-10, but the white marks help them concentrate on spotting the utilities that are near the excavation.

White is used to indicate proposed construction. The chart in Figure 2-11 shows the colors used to indicate each type of utility. A directory of one-call organizations is included in the Appendix.

FIGURE 2-10. Before beginning any excavation activity, call the local one-call organization and request that underground utilities be marked on the site.

Excavating Safely

Once the foundation walls are in place, even a shallow excavation is a *trench,* according to the Occupational Safety and Health Administration (OSHA). To minimize the likelihood that the sides of an excavation will collapse and potentially injure a worker, the dimensions of the excavation must conform to current OSHA standards for the *depth* and *angle of layback*.

For most soils, if the sides of the excavation slope at a ratio of 1-foot vertical to 2-feet horizontal, collapse is unlikely. This slope does not guarantee compliance with OSHA regulations, however. When water is present, either as a result of a high water table or precipitation, the slope may need to be even flatter.

Compliance with the OSHA excavation regulations can be difficult. If the excavation is 6 feet deep and the residence is 5 feet from the property line, the required sloping of the sides of the excavation will extend beyond the property line. In addition, because more earth must be excavated to properly slope the sides of the excavation, the volume of the backfill will increase. The additional backfill must be carefully placed to avoid unnecessary pressure on the basement walls.

NAHB has worked with OSHA to develop a solution for these situations. In 1997 an NAHB educational program, *Construction Safety for the Home Building Industry,* introduced safety requirements adapted for residential construction. This program provides an alternative solution for compliance with OSHA regulations that is designed to protect workers from cave-ins during house and basement foundation construction. These proposals apply when:

- The house foundation or basement is less than 7 1/2 feet deep or is benched for at least 2 feet horizontally for every 5 feet or less of vertical height.
- The width of the excavation is greater than 2 feet.
- The are no water, surface tension cracks, or other environmental conditions present that might reduce the stability of the excavation.
- There is no heavy equipment or vehicular traffic operating in the vicinity that could cause vibrations to the excavation while workers are in the trench.

When these conditions are met, the proposal states that:

- All soil piles, equipment, and materials must be placed no closer to the edge of the excavation than the depth of the excavation. When front-end

FIGURE 2-11. The buried utility lines are marked by color.

The following marking guide provides for universal use and understanding of the temporary marking of subsurface facilities to prevent accidental damage or service interruption by contrators, utility companies or any others working on or near those underground facilities.

Red	— Electric power lines, cables, conduit, and lighting cables
Yellow	— Gas, oil, steam, petroleum, or gaseous materials
Orange	— Communication, alarm or signal lines, cables, or conduit
Blue	— Water, irrigation, and slurry lines
Green	— Sewers and drain lines
Pink	— Temporary survey markings
White	— Proposed excavation

Source: Lara Carr.

loaders are used to dig the excavation, the soil excavated must be placed as far back from the edge of the excavation as possible, but never closer than 2 feet.

- The number of workers in the excavation must be kept to a minimum.
- The work must be planned to minimize the time the workers are exposed in the trench.
- The site must be inspected daily for changing conditions that might affect the stability of the excavation.
- All workers that may be exposed to the potential hazard must receive appropriate safety training.

Additional applicable OSHA regulations that must be followed include:

- A form of egress such as a ladder, ramp, or equivalent must be provided within 25 feet of each worker if the excavation is deeper than 4 feet. If a ladder is used, it must extend a minimum of 3 feet above the excavation and be certified as compliant with OSHA requirements. A job-built ladder is not sufficient.
- A *competent person* must be present at the jobsite when workers are in any trench. A competent person is an individual with knowledge of soil types and hazardous conditions. This individual has the authority to control and stop the work until the hazardous situation is corrected. The person may be the builder, the builder's superintendent, an employee, or, if appropriate, a responsible party employed by a foundation subcontractor.

Pretreating to Prevent Termite Infestations

Many regions of the United States have termites. These insects live in the earth but can enter a residence and feast on lumber and other cellulose material. The foundations of homes built in these areas should be pretreated for termites. The EPA sets standards for the amount and strength of chemical termiticides that can be used in treatment solutions. Most effective termiticides can be purchased and applied only by licensed individuals who have been tested and certified, usually by the state's Department of Agriculture. The types of chemicals used include chlorpyrifos, cypermethrin, fenvalerate, or permethrin sold under brand names such as *Dursban TC, Demon TC, Prevail FT, Tribute,* or *Dragnet FT*.

Builders should be aware that pretreating a foundation with a chemical solution that is too weak could be just as ineffective as no treatment at all. To be sure the solution used is strong enough, and that the customer is getting his or her money's worth, check the quantities of treatment chemicals noted on the work slip. In general, a termiticide should be applied at the rate of 1 gallon per 10 square feet of floor area. In addition to slabs, the cores of blocks in foundation walls and the soil around the perimeter of a residence also may be treated with a termiticide.

Chapter Checklist

When preparing or evaluating a site layout, answer these questions.

- Has the building permit been issued?
- Is activity taking place on the correct lot?
- Have the lot lines been verified?

Before beginning any excavation at the site, ask:

- Are the locations of all underground utilities verified?
- Has all vegetation, including trees and stumps, and topsoil within 20 feet of the planned structure been removed properly?
- Have all trees that are to be saved been identified clearly and protected as effectively as possible from potential damage?
- Are workers trained to follow safety guidelines and practices for excavations? Have all risks been mitigated?
- Are locations for stockpiles of excavated soil identified and clearly marked so that the piles will not interfere with construction traffic?
- Have precautions been planned and implemented to minimize the erosion of excavated soil?

3

Siting the Foundation and Footings

After the site is cleared of vegetation and the topsoil stockpiled, the next step is to mark the perimeter of the residence. Once the perimeter is determined, footings must be located either after excavating the basement or within the slab. Within this chapter are helpful suggestions and formulas for measuring and marking each of the components required in residential construction.

Before the perimeter of the home can be marked a copy of the foundation plan must be obtained. Count the total number of corners in the footings and foundation. A stake will be needed for each corner. The perimeter can be marked using either the diagonal distances between the corners or the 3-4-5 triangle method.

Before laying out the residence, hammer a nail into as many wooden stakes as there are corners in the home. The nails will be used to secure the tape measure when the corners are marked. Some builders secure the tape measure at the corners using screwdrivers inserted through the tape's loop and pushed into the ground to serve as temporary stakes. Another alternative to wooden stakes is 18- to 24-inch-long pieces of 3/8-inch-diameter rebar driven into the ground.

Using Diagonal Distances

Equal diagonal measurements can be used to confirm that the layout is square because the distances between the opposite corners of a square or rectangle are equal. Divide L- and U-shaped residences into a series of rectangles and calculate the diagonal distances between the corners. Figure 3-1 illustrates how to use the diagonal measurements to square the foundation.

FIGURE 3-1. Diagonal measurements can be used to confirm that walls are square.

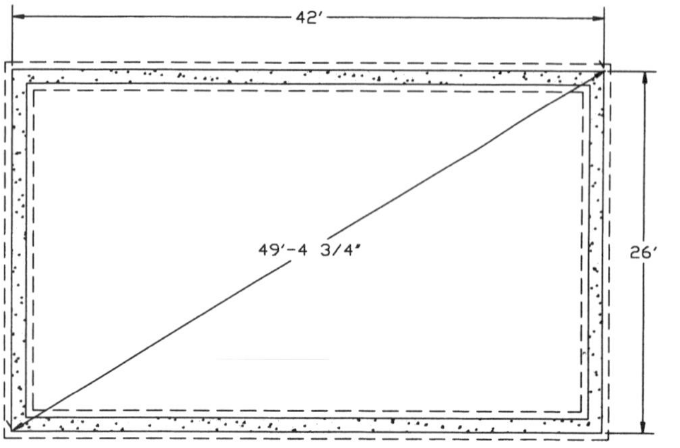

Calculating Diagonal Distances

The diagonal distances between the walls in the foundation can be calculated from the dimensions on the plan using the Pythagorean theorem ($A^2 + B^2 = C^2$) and a hand-held calculator.

If the dimensions of the residence are 42×26 feet, for example, the diagonal distance between the front and back corners is 49 feet, 4 3/4 inches. To calculate this, square both the building's length of 42 feet and its depth of 26 feet and add the two values together.

$$42 \times 42 = 1764 \text{ feet} + 26 \times 26 = 676 \text{ feet} = 2440 \text{ feet}$$

Next find the square root of 2440 feet using the calculator. The square root is 49.396356, which is the diagonal distance between the front and back corners.

Now comes the hard part—how far is this? The number to the left of the decimal point— 49—is the number of feet. The number to the right of the decimal point—.396356—must be converted from decimals of a foot to inches. To do this, multiply the number to the right of the decimal by 12. The result—4.756272—is the number of inches.

$$.396356 \times 12 = 4.756272$$

This answer, however, still has decimals in it—.756272. The last step depends on how precise the measurement needs to be. Usually, 1/8 of an inch is acceptable. To find the nearest 1/8 of an inch multiply the numbers to the right of the decimal—.756272 by 8. The result is 6.050176, or 6/8 of an inch. This can be reduced to 3/4 of an inch. (Remember from fifth grade that all fractions must be reduced.)

Putting together all the calculations, the total is an approximate distance of 49 feet, 4 3/4 inches, which is a distance that can be measured on most tapes. A calculator that accepts dimensions in feet and inches and expresses values in fractions of an inch is helpful. Engineer's tapes that measure in feet and hundredths of feet also can make the task easier. The designer or architect also may have placed the diagonal distance on the foundation plan. In addition, most computer-aided drafting (CAD) software can calculate this dimension. Figure 3-2 is a photo of a foot-inch calculator.

FIGURE 3-2. A foot-inch calculator simplifies the process of multiplying measurements of mixed dimensions, such as feet and inches.

Source: Calculated Industries, Inc., Carson City, NV.

Placing the Stakes

From the strings used to define the lot lines measure the appropriate sideyard and front setback to find one of the front corners of the residence. Mark the first corner by driving one stake firmly into the ground. Secure the loop at the end of a 50- or 100-foot tape measure to the nail. Measure the distance along the front of the home and drive a second stake into the ground to indicate the second corner. Attach a second tape measure to this nail. Then walk to the approximate location of the third corner.

Hold the tapes together so that the distance of the second side of the home on one tape and the diagonal distance on the other overlap. One tape may need to be wound up and the other let out. The point where both tapes are taut is the third corner. Drive in another stake and then use the same system to find the remaining corners. Some builders connect the stakes with string or spray paint a line between the stakes to define the perimeter of the residence.

Using the 3-4-5 Triangle

A second method of squaring the perimeter of the residence uses the 3-4-5 triangle. This process, generally referred to as "squaring the corners," is used to check that the corner stakes are perpendicular to each other after they have been placed in the ground.

Connect the corner stakes with string to represent the walls. From a corner made by the intersection of two perpendicular string lines, measure along one string line a distance of 3 feet. From the same corner measure along the other wall line a distance of 4 feet. The distance between these marks must be 5 feet if the walls are to meet at a right angle. If the measurement is not 5 feet, move one of the lines until the measurement is correct. Do not change the intersecting corner that was the starting point for the wall measurements. If the intersection of the two lines changes, the 3- and 4-foot lengths must be measured again.

It sometimes is simpler to use longer lengths. This procedure works with all multiples of 3, 4, and 5. Measurements of 9-12-15 or 30-40-50 also will verify that the corners are square. Make sure that the tape is read accurately. When working with distances of more than 20 feet, don't let the tape sag or the reading will be incorrect. Keep the tape taut and make sure 6s and 9s are read accurately.

Continue the process by moving counter-clockwise or clockwise around the perimeter of the house. Adjust stakes and string lines until the perimeter is squared. Make sure that corners that have been squared once are not readjusted when checking another corner or the process can be endless.

Restringing for Basement Excavations

For basements, the base of the excavation should be wide enough to provide a minimum of 2 feet of working space around basement walls. Providing a larger space may seem practical, but the weight of the extra backfill against the basement walls increases unwanted stress. Use batter boards or offset the stakes 2 feet to mark the foundation layout and expedite the restringing of lines for footings and foundation walls. The depth of the foundation excavation usually is indicated on the plans.

One worker alone can take advantage of current technology and use a laser level and a receiver attached to a rod to determine the correct depth. Figure 3-3 shows workers using a laser level to determine depth.

Reestablishing the Footprint After Digging the Basement

Run lines from the marks on the batter boards or stakes placed to site the residence on the lot. Drop a plum bob from the intersection of the sting lines down into the excavation to locate the corners. Then run a second string to establish the outside edge of the footing, a distance that generally equals one-quarter of the width from the each corner. This offset is needed because the footing usually is twice as wide as the foundation wall is thick.

Use a second string line to mark the exterior perimeter of the foundation wall. This string line should be attached to a nail that has been driven into the ground. The top of the stake should be as high as the top of the foundation wall. Use a builder's level or laser level to drive all the stakes to the same level. Use the string lines as a guide to place the outside forms.

FIGURE 3-3. A laser level mounted on a tripod is an effective tool for determining the depth of an excavation.

Marking Footings within the Foundation Perimeter

Once the perimeter of the foundation is marked, the footing can be located. Center the exterior edge of the foundation wall on the footing. The offset distance from the edge of the footing can be found in the section drawing in the plans. This drawing describes the foundation and provides dimensions for the foundation wall and the footing.

Because the foundation wall is centered on the foundation the offset distance is equal to half of the difference after subtracting the width of the foundation wall from the width of the footing. For example, if a 20-inch footing supports a 10-inch block stemwall the offset would be 5 inches.

20 inches – 10 inches = 10 inches/2 = 5 inches

Because most footings are twice as wide as the foundation wall this distance usually is one-quarter the width of the footing.

The location of the spot footings required to support point loads such as posts that support girders and interior fireplaces also must be laid out. Usually the distance from the exterior of the foundation wall to the center of the spot footings is indicated on the foundation plan. The dimensions, typically 24×24×12 inches deep, also are shown on the plan. When laying out these footings it is important to remember that the forms for the footings are wider than the foundation walls. This offset must be accounted for when marking the center of the spot footing. The dimensions and the sizes of other spot footings also are marked on the foundation plan.

The top of the spot footing should be level with the top of the footing for the foundation walls, even though the spot footings may be deeper than the footings supporting the foundation walls. It is not necessary to completely form the sides of the spot footings. Most are set with 2×4s and the sides filled in below the base of the lumber to hold the concrete in place.

Aligning the Forms

To ensure that the walls of a basement foundation or that stemwalls are straight, install a string line that is 1/4 of an inch outside the exterior edge of the wall forms. Attach one end of a string line to a nail or piece of rebar at one corner, run the line along the wall to the next corner, and then attach it to another rebar piece. The forms should be 1/4 inch inside the string line. Push or pull any forms that are out of alignment to make sure that the cast wall will be straight and plumb. When all the forms are 1/4 of an inch inside the length of the string line, secure the rakers to hold the forms in place. Workers using a string line to align the forms are shown in Figure 3-4.

FIGURE 3-4. A string line located 1/4 inch from the true outside edge of the wall makes it easier to align the forms.

Source: Courtesy Fold Form LC, Sioux City, IA.

Chapter Checklist

When marking the perimeter or setting up string lines to align the forms, make sure that the following tasks are complete.

- Are the property corners marked correctly?
- Does the layout correspond to the plot plan?
- Have the zoning setbacks and easements been checked?
- Does the placement of the structure on the lot contour provide for adequate drainage?
- Are reference points located outside the perimeter so that string lines can be established after excavation disturbs the initial markings?

4

Planning for Water Control

Building foundations on soils that are poorly drained or that retain water can lead to problems regardless of the foundation type. An effective drainage system and a foundation designed for the site can help minimize future problems.

Look carefully at the building site. What kinds of vegetation occur naturally on the site? Cattails, marsh grass, and willow trees can suggest the presence of water close to the surface. Test borings at the site also can indicate potential water hazards. If a test boring is made 2 to 4 feet deeper than the depth of the proposed foundation and water is evident at the base of the boring after 24 hours, it is best to anticipate water problems.

The topography of the site also can cause water problems. If the building site is in a flood-prone area, it must be assumed that at some point the site will flood. The site need not be near a river for this potential to exist. A heavy rain can create water problems for sites located in a depression below only a few acres of watershed or in a channel that might be an intermittent streambed.

The most visible sources of problem-causing water in residential construction are rain and snow. Other sources, such as groundwater, may not be visible at all. Improperly graded yards, a lack of gutters, or the presence of cracks in the foundation wall provide visible, and easily prevented, access points for unwanted water and moisture.

Because moisture can enter at one point and travel several feet before causing a visible problem entry points can be hard to detect. Once the house is complete, finding and eliminating the routes taken by troublesome moisture becomes more difficult and expensive to solve.

Homeowners do not want wet or even damp basements. Although their water problems may not be as obvious, owners of homes with a crawl space or slab foundation also can face serious water-related problems. Moisture that is present in, under, or around any structure, for example, may promote the decay of the floor joists. At the least, mold and mildew thrive in the house's moist environment creating an undesirable environment.

No one preventive technique guarantees a dry basement or a foundation free of moisture problems. In addition to footing drains, the use of a bituminous coating and a polyvinylchloride (PVC) film or a cold-applied elastomeric coating and polystyrene drainage mats may be required. Many systems only can be applied by licensed or franchised operators and may not be available in some areas.

Cost also can be a big factor. Some of these systems require a lot of labor and most require skilled workers. Trying to cut costs by cutting back on the skill level of laborers can backfire. To work correctly water-control and damp-proofing systems must be installed correctly the first time.

Developing a Drainage Plan

The best way to keep any structure dry is to keep water from coming in contact with it. To move water away from the home, each plot plan should include a drainage plan for the finished residence. The finished grade around the perimeter of the structure should fall away from the dwelling at a minimum slope of 6 inches for the first 6 feet. After 6 feet the slope can be reduced to a minimum of 2 feet for every 100 feet.

Installing gutters, downspouts, and splash blocks that direct rainwater away from the residence helps control the removal of rainwater. A good overall landscaping plan helps eliminate the possibility that water will be trapped near the structure. Figure 4-1 is a drainage plan drawing.

FIGURE 4-1. A drainage plan drawing includes the structural elements required to keep rainwater away from the residence, such as gutters and downspouts.

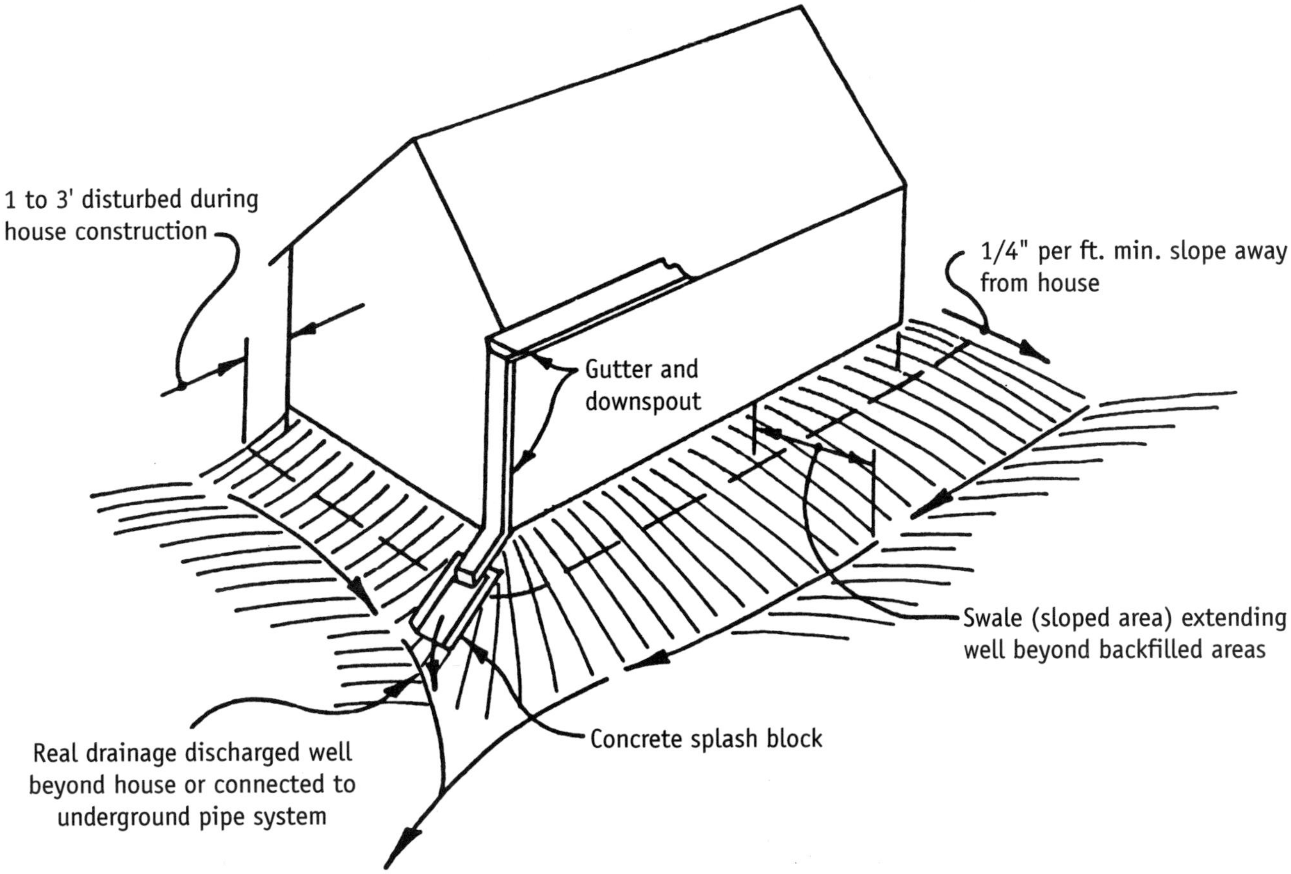

Source: Residential Drainage, *by S. Spano and D Isokait, Washington Suburban Sanitary Commission, Hyattsville, MD; 1979.*

Homeowners can unknowingly trap water in the landscaping around their residence. Wood, metal, and other types of edging can form a barrier that prevents water from moving away from the perimeter of the home. Layers of mulch more than 2 inches deep also provide a place for moisture to collect. Educating homeowners on the importance of maintaining drainage paths away from the home can help prevent future problems.

Understanding System Types

Building codes require a damp-proofing system that retards the migration of moisture through a basement wall when the backfill around the perimeter of the basement is not saturated with water and hydrostatic pressure does not exist. If groundwater in the soil around the basement adds hydrostatic pressure a waterproofing system is required. Making the wrong decision about which technique to use can lead to costly repairs. Usually a system that decreases the hydrostatic pressure on the foundation wall and a system that prevents moisture from moving through the walls should be provided.

A *dewatering* system places pipes in the ground to collect water and direct it away from areas where it might create problems. These systems include french drains and footing drainage systems. A number of drainage systems and materials are available. Most drainpipes are laid in a gravel bed. The gravel filters some of the silt from the water and increases the efficiency of the drainage system.

Providing Site Drainage

In the past hollow round clay tiles were used as drainage pipes. In some regions of the country drainage pipe still is referred to as tiles. The most common drainage pipe used for this purpose is either a smooth-rigid or flexible-corrugated plastic pipe. Parts of the pipe can be perforated with slits or holes to allow water to enter. Perforated pipe also can be connected to nonperforated sections designed to simply move the already-collected water. Install the perforations with the holes at 2 and 4 o'clock to allow water to flow along the solid bottom of the pipe.

Another term associated with site drainage is *french drain*. This type of drain does not use pipes, but a trench partially filled with rock or gravel. French drains depend on the gravel not clogging up with silt. Depending on soil conditions, the trench sometimes is topped with a geosynthetic material that traps the silt in groundwater but does not impede the flow of water into the trench. Adding corrugated drainpipe to help keep the system from clogging also can be a good investment.

Including Footing Drains

If the foundation includes a basement or if the house, regardless of the foundation used, is located in an area with a high water table, a footing drain is a required part of the drainage system. These drains make the whole water-control system more effective by moving water out of the backfill and reducing the hydrostatic pressure on the walls. The most common way to create this footing is to place a 4-inch-diameter perforated pipe along the interior side of the footing. Place the pipe so that the bottom is above the base of the fill under the basement slab and no closer than 12 inches from the edge of the footing. In extremely wet conditions, pipe also might be required on the exterior of the foundation. Drainage pipe for this system is available in rigid sections that can be joined to elbows or in rolls of corrugated piping that can be curved around corners.

Placing the pipe is easier if the footing is cast in a form rather than in a trench. The drainpipes should fall at least 1 inch every 10 feet. Place the pipes on a layer of gravel or crushed stone at least 2 inches deep and cover them with 2 to 4 inches of gravel. The gravel should be a mix averaging 1 1/2 inches in diameter washed free of dust that can clog the system.

The top of the gravel bed also may be covered with a filter fabric. The fabric prevents silt in the groundwater from seeping into the gravel bed, increasing the effectiveness of the system. Geotextiles developed for this purpose are available in rolls from many building or landscaping supply companies. Because the geotextile does not prevent the movement of moisture, a vapor barrier still is required under the basement floor. Stay-in-place forming products also are available that automatically incorporate the drainpipes, saving time and labor.

Placing the Gravel

In some areas of the country trucks with conveyor belts are available to facilitate the placement of the gravel. Gravel is delivered to the site and dumped into a hopper. The gravel then is moved along a series of conveyor belts. Some of the belts are attached to a boom that can rotate and extend. At the end of the boom is a second hopper into which the gravel falls. At the bottom of the hopper is a hose that the gravel flows through. An operator controls the speed of the conveyors and the positioning of the end of the boom. Some additional leveling of the fill may be required but this system greatly reduces the time this step takes. A section of a foundation drainage system is shown in Figure 4-2.

Installing Sump Pits

Regardless of the foundation type, the perforated drainpipes collect and convey the water to *daylight* or to a *sump pit*. Daylight is the term for systems that allow water to flow until the drain is exposed and the water runs onto the surface of the ground. The water that collects in a sump pit is pumped away mechanically.

Prefabricated plastic or cast concrete sump pit boxes are available. Most sump pits are about 16 inches in diameter and 24 inches deep. The sump boxes and all the drainage pipes are placed under the slab and cast into place so that the top of the sump box is flush with the floor. Sump pits provide knockouts through which the drains run and a lid with a hole through which the pump's pressure pipe and electrical cord run.

Because the sump pit drainage system relies on the operation of the pump, the pump should be checked at least once a year. If the pump pit is covered and sealed, which sometimes is done to protect against radon gas infiltration, reseal the lid after inspection. If possible, connect the pump to a backup battery system that can sense when the power to the house is cut off. The backup battery should be able to operate the pump until the power is restored.

Sensors activate the sump pump by sensing when water in the sump pit reaches a predetermined level. Most pumps are submersible and work by sensing the water pressure against a diaphragm. In others, the level is adjusted with mechanical floats. The sump pump is connected to an outlet pipe, usually a 1 1/2- inch-diameter pipe. The outlet pipe carries the water from the sump box to a disposal point. The pipe leaving the pump should be flexible or have connections that allow for the pipe to be removed from the pump for repair or replacement.

The water in this pipe can be discharged onto a splash block on the surface or the outlet pipe can run below the surface of the yard to a discharge point away from the home. Many municipalities do not allow water drained from a sump pit to be connected to the

FIGURE 4-2. The foundation drainage system illustrates the drainage elements incorporated into the foundation, such as the sump pump and the foundation system connection through the footing.

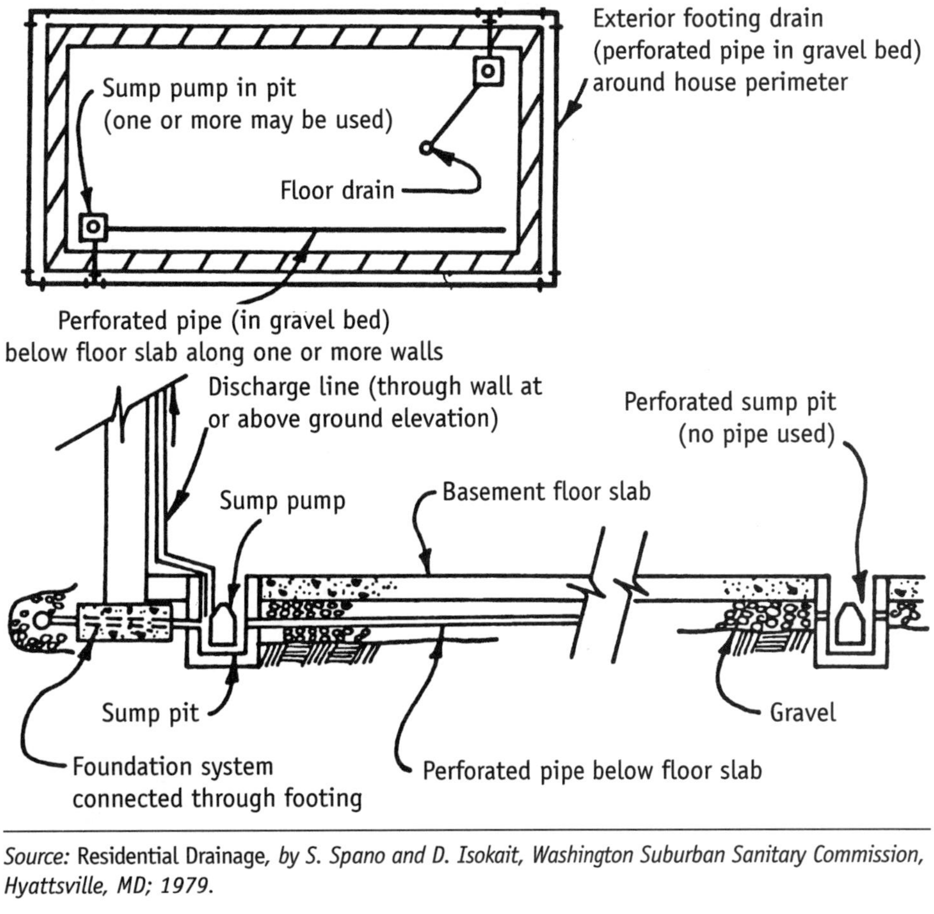

Source: Residential Drainage, *by S. Spano and D. Isokait, Washington Suburban Sanitary Commission, Hyattsville, MD; 1979.*

sanitary sewer system. This practice might overload the capacity of the sewage treatment facilities.

Placing Drainage Pipe

Most builders prefer to place one drainage pipe under the basement floor along the exterior walls. The justification for this is that if the foundation walls leak and the subgrade under the footing leaks, or if water rises up under the residence, a drainage system located there will capture the water and direct it away from the structure. The integrity of the drainage pipes placed along the exterior side of the wall, however, can be diminished if they are damaged during the placement of the backfill.

If drainage pipes are placed along the exterior of a foundation wall they should be placed on a 2- to 3-inch-deep and 12-inch-wide bed of gravel and covered by at least 6 inches of gravel. Place the gravel and the first 2 feet of backfill carefully. Do not compact the backfill to protect the drainage pipe from damage.

Creating Drainage Around Block Walls

Some builders intentionally construct concrete block basement walls to allow water to flow into the interior drainage system. Walls built with notched concrete blocks or corrugated plastic or metal plates placed in the mortar between the footing and interior side of the first course allow water that has collected in the cores of the block wall to flow from the base of the wall to the drainage system.

Water-control or damp-proofing measures are applied to the exterior of the walls to minimize water penetration. The notches, slots, or other measures, however, ensure that water that does penetrate can move through the wall in a controlled manner and collect in an underfloor drain. For this type of system to succeed, the depth of the underslab drainage bed must be adequate. The top of the fill should be 2 inches above the footing on the inside of the foundation wall. This gives moisture passing through the wall an avenue to the drain. The block drainage system is illustrated in Figure 4-3.

Installing Integrated Footing and Drainage Forms

Foundation drainage, with drainage pipes placed near the base of the formed footing, is part of the damp-proofing or waterproofing system for most foundation plans. Two alternatives to this system are available. The key benefit to these alternatives is that the plastic drainage pipes serve as the form for the footing and remain in place after the concrete sets to form the perimeter drainage system.

An additional benefit to both alternative drainage systems is that they can be incorporated into an effective passive radon gas venting system. This is accomplished by installing an outlet coupler with a connection to a horizontal pipe running through the gravel subgrade under the basement slab. Connect the pipe with a Tee to a vertical stack that vents through the roof of the residence. The Tee connection vents the area under the slab and the drainage pipes incorporated into the walls vent the perimeter of the basement. The vertical stack is warmed by the heat of the residence and promotes a

FIGURE 4-3. Concrete basement walls are built so that water flows into footing and underfloor drains. This material also serves as a slip sheet, reducing shrinkage cracks by preventing the concrete floor from binding to the foundation wall.

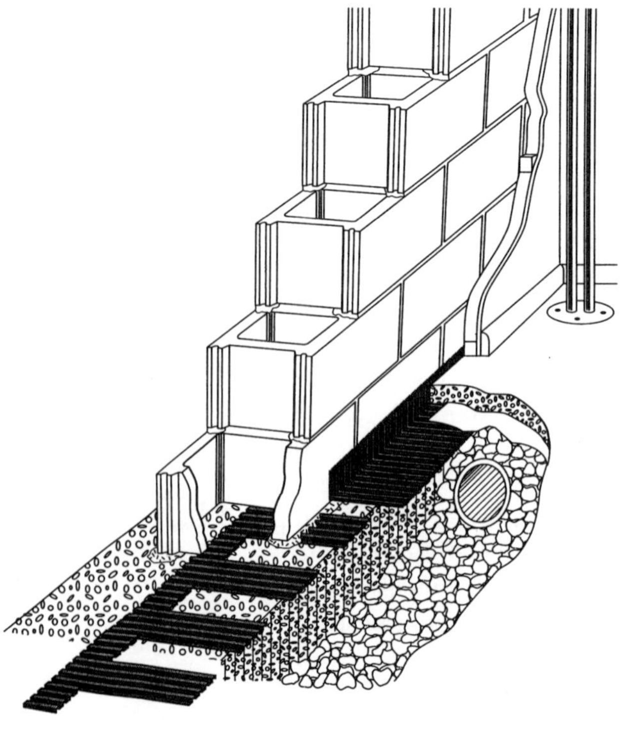

Source: Masonry Technology Incorporated, Cresco, IA.

chimney effect to power the passive system. The passive vent system is illustrated in Figure 4-4.

Installing Rectangular Plastic Channels

One system uses specially designed rectangular plastic channels to meet the drainage requirements of foundations. The lightweight, rectangular PVC forms are available in 12-foot lengths that are 4, 6, or 8 inches tall and 2 inches thick. These plastic forms, which include special extruded corners, are set in place using techniques similar to those for lumber forms. The *lineals* or rectangular sections are manufactured with one solid and one slotted side to allow water to flow into the drainage system. Hollow corners provide for continuous drainage. Special adapters permit the system to connect to round drainage pipe that then can connect the system to the sump or drain to daylight.

To install the plastic channels, set a string line that is offset by 2 inches along the exterior side of the footing. Place a lineal section of the PVC form with a corner joint fitted on one end at a corner. Hold the section in place by driving a rebar dowel into the ground through the hole manufactured into the corner section. Then slide the next lineal section with a coupling joint into place. Drive another rebar dowel through the hole in the top and bottom of the coupler to align the form with the string line. Continue the process until the pipe is placed along the outside of the footing. The lineal sections can be cut with a handsaw. Couplings are available for straight lines, 90- and 45-degree angles, and Tees and Ls for stepped footings. Figure 4-5 shows the drainage system with installed Tees.

After the exterior side is installed place the interior sections. Align the interior side with the installed exterior side using metal spacers sized for the various footing widths. Level the forms by driving metal or wooden stakes about every 5 feet along the length of the forms and securing them with drywall screws or duplex nails. Drive additional drywall screws partially into the interior side on the forms that will be cast into the footing to serve as an anchor to hold the drains snug against the footing. If a void is left under the form that could let concrete ooze out, fill it with gravel.

FIGURE 4-4. A perimeter drainage system can be incorporated into an effective passive radon gas venting system.

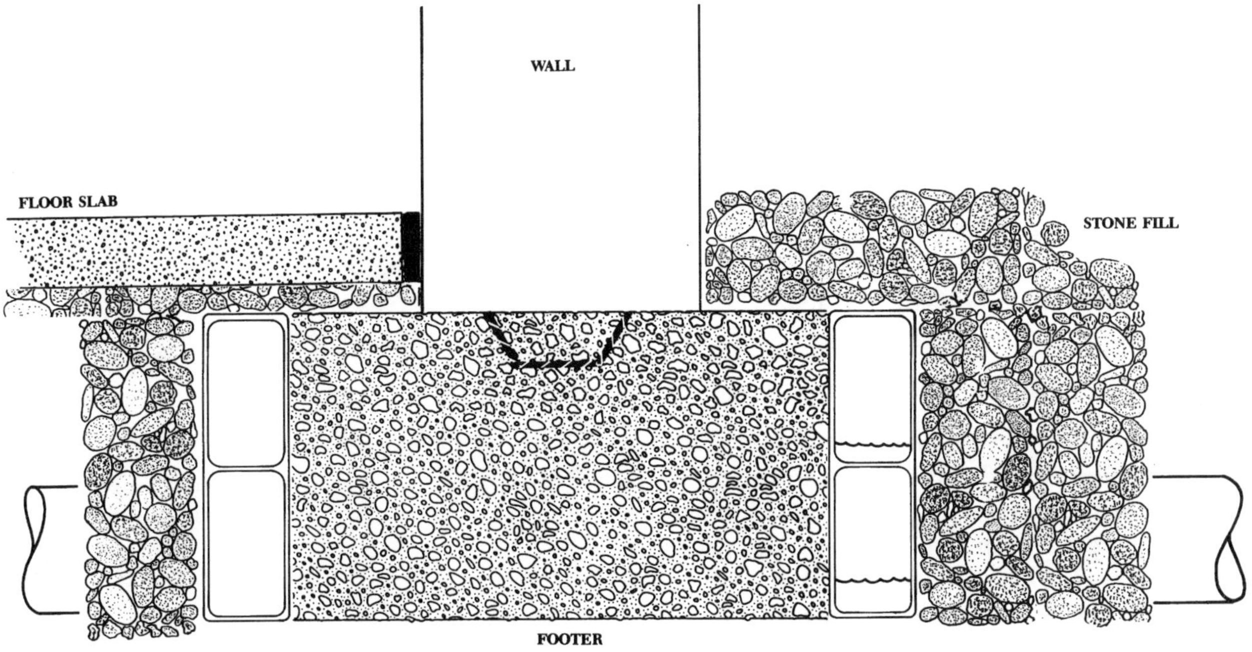

Source: Certainteed, Valley Forge, PA.

FIGURE 4-5. Tee connectors can be used to connect interior and exterior drainage systems.

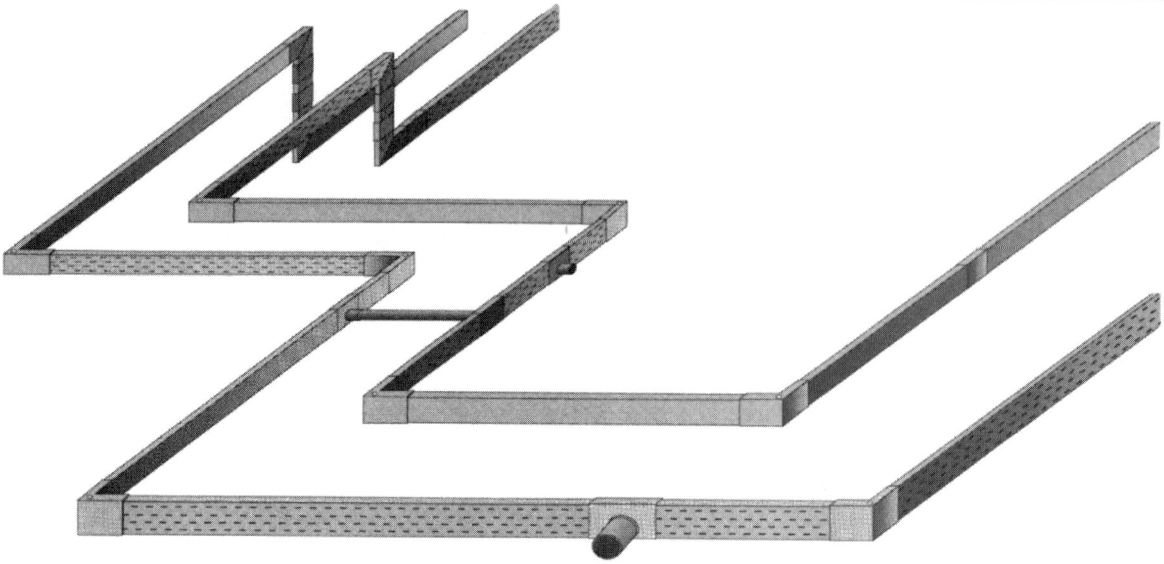

Source: Certainteed, Valley Forge, PA.

Installing Round Perforated Drainage Pipe

Another drainage system incorporates round perforated drainage pipe in the footing form. The key to this system is a unique stake that is placed through a hole in the pipe and then into the soil. The threaded end of the clamp on the stake allows the installer to easily adjust the pipe up and down to level it to the top of the footing's form. An additional smaller diameter pipe connects through the footing to combine the drains into one system.

After the footing system is installed and leveled gravel is placed along the sides of the pipe to complete the "form" and hold the concrete in place. The clamps are removed after the concrete is placed. This system eliminates the need to strip the forms because the drainage pipe is already in place. Figure 4-6 shows this drainage pipe system installed.

Installing a Vapor Barrier

Drainage must be provided under the slab as well as around the perimeter of the foundation. When building in well-drained soils such as those containing a high percentage of sand, the drainage occurs naturally. In areas where the soil has a high clay content, drainage must be provided by placing a bed of compacted sand or gravel at least 6 inches deep under the slab. Gravel works best. Its larger grain size provides larger voids through which water may flow so that it does not pool and penetrate the subgrade.

In addition to the drainage layer, a vapor barrier of either 6-mil polyethylene or 4-mil PVC sheets must be placed under the slab. Overlap the sheets a minimum of 12 inches and repair any tears in the sheets with tape. This layer can be laid down just before the concrete is placed.

FIGURE 4-6. Round perforated drainage pipe can be built into the footing system.

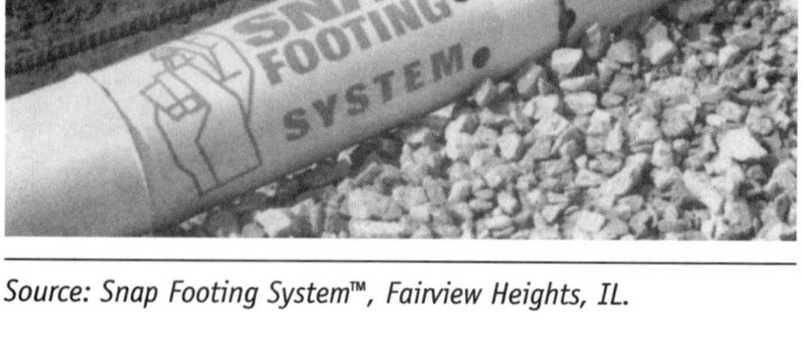

Source: Snap Footing System™, Fairview Heights, IL.

Placing concrete for a slab directly on the vapor barrier prevents moisture from the freshly placed concrete from moving downward into the subgrade. This can lead to difficulties such as bleed water during the curing process and side effects such as hairline cracking and a white dust on the surface. To reduce the likelihood of bleed water, place the vapor barrier under 2 inches of fill. The fill between the vapor barrier and the concrete captures the moisture and provides a route for it to escape.

Understanding Water-Control Products

Several types of coating materials can be applied to foundation walls to help prevent or reduce water penetration. Many of these products are brushed, painted, or sprayed directly onto the exterior below-grade surface of the wall. Workers using these products should wear clothing that protects the body, including rubber gloves.

Every foundation wall must be damp-proofed and treated for water control. Cast concrete walls, although solid, still allow water to flow through them, masonry block walls are very porous, preserved wood walls have nail penetrations and joints that must be sealed, and insulating forming systems have joints through which water can penetrate.

To ensure proper performance, follow the manufacturers' recommendations carefully when installing any water-control product. For example, a cast concrete wall must be cured for 10 days prior to the placement of certain coating materials. If the coating is applied too soon, too much moisture may be trapped within the curing concrete. This moisture could condense on the surface of the foundation wall and create blisters in the water-control compounds. If the blisters break the water-control system will fail. Some products require coating to a specific thickness or application using special tools. Other products must cure for a specific amount of time before any backfill can be placed.

Care should be taken with any water-control product to ensure that the coating is complete. One area that requires special attention is the intersection of the foundation wall and the footing. Placing a coating of cementitious material to round out the transition between the horizontal and vertical surface helps. Inspecting such areas carefully and making an extra pass or two with the trowel, sprayer, or brush as needed can help ensure an unbroken coating in hard-to-cover areas. It is more cost-effective to plan for adequate time and materials to ensure a good seal than it is to reexcavate and repair a poorly done job.

Some manufacturers have specific steps or requirements that must be followed before their product is used. For example, before most products can be used on concrete block foundation walls, the walls must be coated with a 1/2-inch-thick parge coat. Other products can be applied directly to the block without the parge coat. Check the instructions for the product that will be used.

Some products, such as insulating concrete forms (ICFs), contain polystyrene and therefore are incompatible with certain water-resistant coatings. These waterproofing products will dissolve the foam used in these foundation systems. The label on the water-control product should list items with which it is not compatible. To be sure, test a variety of coatings on the building materials to verify product compatibility before purchasing material in bulk.

Some water-control systems require, in addition to materials applied to the foundation wall, a sheet of 6-mil polyethylene or 4-mil PVC. This provides a complete system that protects the residence. Even though these products are available in large rolls some joints will occur. Overlap all joints at least 6 inches and seal them with an adhesive, or the water-control compound. Seal all penetrations, such as pipes. In addition, a flashing usually is required at the top to decrease the chance of water getting behind the plastic sheet.

Applying Bituminous Coating Systems

Bituminous coating systems are among the oldest and most common water-control products. They also are among the least effective. These products, derived from either coal tar or asphalt, are either cold- or hot-applied. Cold-applied products bond well to most clean dry surfaces and usually are applied with a trowel. Hot-applied products have strong odors and are mopped on. They can be difficult to apply consistently on the vertical surfaces of foundation walls.

With age some of these products become brittle. Also, bituminous coatings perform poorly if cracks form in the foundation wall. When the wall cracks so does the coating. If supplementary sheeting is applied with this system it should be placed while the bitumen is still tacky, which can be a messy task. If hot-applied bitumen is used, application of the sheeting can be particularly difficult.

Using Liquid-Applied Elastomeric Products

More effective products are made from polyurethanes, epoxies, acrylics, and silicones used alone or in compounds. These liquid-applied elastomeric products can be applied with a brush or spray, although some do require a hot application. Elastomeric products bond well to most clean vertical surfaces. They remain flexible after they have cured and have the ability to bridge over small cracks that may develop. They also conform well to a variety of shapes and can seal tight against obstructions such as pipes passing through the foundation wall. Once torn or scratched, however, most cold-applied products do not self-seal. Hot-applied products will seal surface penetrations up to 1/4-inch wide. In addition most hot-applied products can be applied even in subfreezing temperatures.

Solvent-based liquid-applied elastomeric compounds can emit toxic and flammable fumes. Safety precautions must be taken during application. Safer acrylic water-based compounds cannot be used at temperatures near freezing. All of these types of water-control compounds must be used quickly. Once the container is opened they generally remain workable only for a specific amount of time, depending on the temperature. Also, solvent-based compounds generally cannot be used on cast concrete foundations until the concrete has cured for 28 days.

Applying Cement-Based Products

Another series of common water-control products uses a cement base. These products typically combine portland cement, sand, and various admixtures to create a layer impervious to the seepage of water. A common admixture is ironite, which is mixed with portland cement and sand. A 1/2-inch-thick layer of this product is applied to the wall with a trowel. The ironite oxidizes and expands allowing the parge coat to seal small cracks that may develop. These cementitious products are applied to the foundation surface using trowels, rollers, or brushes. As the material cures, crystals form and penetrate the surface and tightly bond the product to the foundation wall enabling it to withstand significant hydrostatic pressure. To be effective, most cement-based systems require several coats.

Cementitious water-control products are easy to apply, even to freshly placed walls. Water penetrating the product activates some of the noncrystallized material, automatically sealing small cracks that may form. When the coating is applied to the interior surface of the wall, any cracks that are too large to be self-repairing are easily spotted and repaired. A cementitious damp-proofing product being applied is shown in Figure 4-7.

Incorporating Drainage Mats

Drainage mats reduce the hydrostatic pressure on the surface of the foundation wall by directing water that seeps through the soil along the wall to the footing drain. Available systems are made with a variety of materials, including rigid extruded plastic panels, boards of expanded polystyrene, blankets of nylon filaments, or fiberglass blankets and boards. Each system creates a vertical void along the exterior of the foundation wall. Moisture from the backfill enters a cavity in the drainage mat through a series of holes or slits and then falls down into and is collected by the footing drain, which may be separate or incorporated into the mat.

The drain then either channels the water to daylight or is connected to a sump pit. Drainage mats

FIGURE 4-7. Cementitious damp-proofing products automatically seal small cracks.

should be used in conjunction with water-control products applied to the foundation wall. Some drainage mats also have value as thermal insulation. A Tuff-N-Dri® mat is shown in Figure 4-8. An example of the Delta MS mat is shown in Figure 4-9.

Additional information on controlling water in residential construction can be found in *Residential Water Problems: Prevention and Solutions* by Alvin M. Sacks, Home Builder Press, Washington, D.C.; 1994.

Chapter Checklist

Consider the following items when installing a water-control system for a foundation.

- Do site conditions such as a high water table or poor soil drainage suggest one foundation type over another?
- Will the slope of the site allow for adequate surface drainage?
- Will the structure include gutters and other features to help direct precipitation away from the foundation after construction is completed?

FIGURE 4-8. Drainage mats, such as this Tuff-N-Dri® mat, reduce the hydrostatic pressure on the surface of the foundation wall.

Source: Koch Waterproofing Solutions, makers of Tuff-N-Dri® and Watchdog Waterproofing®, Toledo, OH.

FIGURE 4-9. Installed against the exterior of the foundation wall, this Delta MS mat, like other drainage mats, channels moisture from the backfill to the footing drain.

Source: Cosella-Dorken Products, Inc., Beamsville, Ontario, Canada.

- How well will the planned system perform on this particular site?
- Does the footing design allow for easy installation of a footing drain? If a drain will be used, the footing should be formed, not cast in a trench.
- Can the footing drain be run to daylight or will a sump pit and pump be needed?
- Will the wall require a parge coating?
- If a chemical water-control compound will be applied, is the coating compatible with the other materials in the foundation wall?
- Have all protective coatings and materials been installed carefully, thoroughly, and according to manufacturers' instructions?
- Is the gravel bed sufficiently large?
- Is the filter fabric in place before the backfill is placed?
- Is the vapor barrier under the basement floor properly installed by overlapping pieces 12 inches and taping tears?
- Can the backfill be placed without destroying the water-control system?

5

Working with Concrete

Concrete is a very versatile building material. It is economical, durable, and very strong. Concrete generally is described in terms of its *compressive strength* or the amount of load it can carry per square inch. Common compressive strength values for residential construction are from 2,000 to 4,000 pounds per square inch (psi).

Reinforced concrete is made with steel reinforcing bars, which add *flexural* strength, or resistance to bending, to the compressive strength of the concrete. Even if codes do not require rebar, adding reinforcing steel to concrete can reduce the thickness of a foundation wall. In regions where concrete is expensive it may be more economical to use steel reinforcing bars to reduce the amount of concrete needed for the job. When calculating the difference, be sure to include both labor and material costs to get an accurate estimate.

Because concrete can be formed into various shapes and finished with a wide variety of textures and colors, builders can employ a number of techniques to create basement foundations, crawl spaces, or slabs on grade.

Considering Relevant Factors

Several factors must be considered when planning a cast-in-place concrete footing or foundation. The availability of concrete is the first. In most areas concrete is delivered to the jobsite in ready-mix trucks from a nearby batch plant. Concrete is plastic for a finite amount of time. Once it begins to harden and set up it cannot be placed into forms or footing trenches. The bottom line is that delivery over distances approaching 100 miles can make the use of concrete uneconomical.

As an alternative, water can be added to the dry materials at the jobsite or during transit to the site. Care must be taken however, to add the appropriate amount of water. Adding more water than that specified in the design of the mix significantly reduces the final strength of the concrete.

The second consideration also involves materials. Cast-in-place walls require forms, which can represent a significant financial investment. In some areas concrete contractors specialize in casting foundation walls and maintain an assortment of forms. In most regions the forms can be rented for a short time. Some contractors use site-built forms and then, after the foundation walls are cast, reuse the lumber and plywood from the forms in the construction of the residence, although some owners and building officials do not like this practice.

The third consideration is labor. Building and erecting the forms correctly requires skilled workers. Manufactured forms do not require carpentry skills but improper connections or lack of attention to details when placing the forms or bracing can create a substandard product. Improperly braced walls can lead to blowouts that waste material and also create a mess that must be cleaned up before the concrete sets. If the walls are not aligned properly, the framing plan may need to be revised. The use of skilled craftspeople, knowledgeable in the cast-in-place process, is essential.

The American Concrete Institute (ACI) has a certification program for concrete finishers. To obtain certification an individual must be able to certify that he or she is an experienced finisher with the expertise to properly place and finish concrete. Those seeking certification must meet specific requirements and demonstrate proficiency. The ACI offers training programs for individuals interested in certification.

Weather is a fourth consideration. Concrete can not be cast in subfreezing temperatures without a source of heat to prevent the freshly cast footing, floor, or wall from freezing. If the concrete freezes within two days of being placed, it will not gain its designed strength and may potentially fail. While concrete is curing, the heat from hydration produced by the chemical reaction between the cement and water will, if retained in the wall, protect the newly cast concrete from freezing if the ambient temperature is at or just below the freezing point. Insulting blankets or insulating forms may be necessary to retain this heat and prevent the wall from freezing. These products, however, increase the cost of the job.

Concrete also is affected by daily fluctuations in temperature. Temperatures that change from above freezing to below freezing stress the concrete. In below-freezing temperatures moisture inside the concrete freezes and expands. Then as the concrete warms, the water melts and shrinks. This process is referred to as the *freeze-thaw cycle*. The pressure exerted during the process can break up or *weather* the concrete. *Air entrained* concrete, in which thousands of air bubbles have been introduced, should be used when the concrete will be exposed to weathering. A higher strength concrete also should be used when the potential for severe weathering exists.

Specifying the Concrete

Most trucks carry between 8 and 10 cubic yards of concrete. When placing a 4-inch-thick slab, a yard of concrete will cover 80 square feet. Most concrete companies schedule the delivery so one truck is emptying the final part of its load as the next truck arrives. Many trucks are radio equipped, which allows constant contact with the driver enabling the job supervisor to add or subtract from the last load, referred to as a *check load*, to compensate for errors or areas that were compacted incorrectly. Figure 5-1 shows the placement of concrete from a ready-mix truck.

Most concrete suppliers have standard mixes for basements and other concrete applications in their

FIGURE 5-1. Placement of concrete from a ready-mix truck.

region. The concrete placed in cast-in-place foundation walls should be a minimum of 3,000 psi, or a five-sack mix, which refers to the amount of cement per yard in the batch. In regions where the temperature seldom gets below freezing or when the concrete in the foundation wall will be protected from freezing by insulation behind a masonry veneer, for example, 2,500 psi concrete may be used. The table in Figure 5-2 shows these values.

Air-entrained concrete is produced with an *admixture* that introduces millions of microscopic bubbles into the concrete. The bubbles provide a void into which ice crystals can expand as the moisture in the concrete freezes, effectively increasing the durability of the concrete. The admixture can be introduced into the batch at the concrete plant or added to the portland cement used to make the concrete. The air content should be between 5 and 7 percent.

Mixing in Admixtures

Admixtures such as superplasticizers and water reducers can be mixed in with the concrete to increase the slump without weakening the concrete. Because this can make the job go smoother, the labor saved may be worth the additional expense.

Admixtures also make it possible to transport concrete longer distances by retarding the chemical reactions in the concrete and lengthening the time before set up occurs. Other admixtures, such as accelerators, can be used during cool weather to shorten the set-up time of the concrete.

Calcium chloride is a common accelerator. Up to two pounds of calcium chloride can be added per 100 pounds of cement, creating a 2 percent mix. The calcium chloride increases the initial strength over the first three days of curing. Because calcium chloride promotes corrosion, it should not be used when reinforcing steel is present.

Checking the Slump

Before casting, test each batch of concrete to be sure it has the proper *slump*. Slump is a measurement of the concrete's consistency, which is directly affected by the water content. The less water in the mix, the stiffer the concrete and the less it will flow.

Slump is determined by filling a truncated cone designed for slump tests and then placing it upside down on a flat surface. To properly test the concrete, fill the cone one-third full of concrete and rod it 25 times. Then fill the cone two-thirds full and rod it 25 times. Finally, fill the cone and again rod it 25 times. Turn the cone upside down on a flat surface and lift off the cone. Measure the distance between the top of the cone and the mound of concrete. This distance is the slump. A slump test is illustrated in Figure 5-3.

FIGURE 5-2. Site factors determine the pounds-per-square-inch requirements of the concrete.

Seismic Zone	Height of Unbalanced Fill	Distance Between Interior Supporting	Minimum Wall Thickness	Horizontal Reinforcement In Upper 12 Inches of Wall	Vertical Reinforcement Size and Spacing
0, 1, 2	6 feet	Unlimited	6 inches	Required only	None Required
	7 feet		8 inches		
	8 feet		10 inches		
3, 4	<4 feet		8 inches		
	4 to 8 feet	Less than 8 feet	8 inches	2 #3	#3 @ 18 inches on Center
	<8 feet	8 feet to 10 feet	8 inches	2 #4	#3 @ 18 inches on Center
	>8 feet	10 feet to 12 feet	8 inches	2 #5	#3 @ 18 inches on Center

Source: Compiled from Council of American Building Officials (95) Tables R-304.3a & b.

Proper slump is important to ensure maximum control over the concrete during casting. Most residential concrete, however, is placed without performing slump tests. This can lead to an improper placement and later problems.

Many workers prefer mixes that are a little wet because they require less effort to place. Concrete mixtures at or near a 4-inch slump, however, are easy to place and will be stronger, more durable, and less likely to crack or shrink than concrete that is "wetted up a bit" at the jobsite. Higher slump mixtures, 5-inch and above, exert greater pressure on the forms.

Most snap ties are designed for use with concrete mixes that have a 4- to 6-inch slump. Admixtures, with the exception of air entrainment, can affect slump so it is important that the concrete supplier know the kind of form ties that will be used so that a compatible mix can be provided.

Exercising Care

Safety precautions should be followed at the jobsite, especially around ready-mix trucks. The operator has limited visibility and, when extended, the chute can stick out 20 feet or more. A small sideways movement at the front of the truck can be greatly exaggerated at the end of the chute, accidentally damaging property or hitting a worker. Only the truck's operator should be authorized to add and remove chutes because the hinged joints are extremely hazardous.

Wet concrete can cause skin burns or cement dermatitis. Workers should protect themselves from exposure to wet concrete by wearing protective clothing, rubber gloves, and rubber boots when pouring and finishing concrete. Anytime concrete comes in contact with bare skin, the skin should be washed immediately with soap and water. Workers also should wear eye protection and respiratory devices when performing activities that generate airborne crystalline silica dust, including abrasive sandblasting, concrete or rock drilling, grinding, or cutting. When performing dusty activities such as working with concrete mixes, paper dust masks may be worn. If dust enters an eye, wash the eye with clean water and consult a doctor. Do not rub the eye.

Placing the Concrete

When placing the concrete, work at a pace that allows good control of the concrete. The speed of placement and the weight of the wet concrete exert

FIGURE 5-3. a. The first step in a slump test is to fill the cone in three layers of equal volume, rodding each layer 25 times. b. For step 2, strike off the top and then remove the cone slowly with an even motion, taking from 5 to 12 seconds. Do not jar the mixture or tilt the cone in the process. c. The final step involves measuring the slump with a tamping rod and ruler. The measurement is the concrete's slump.

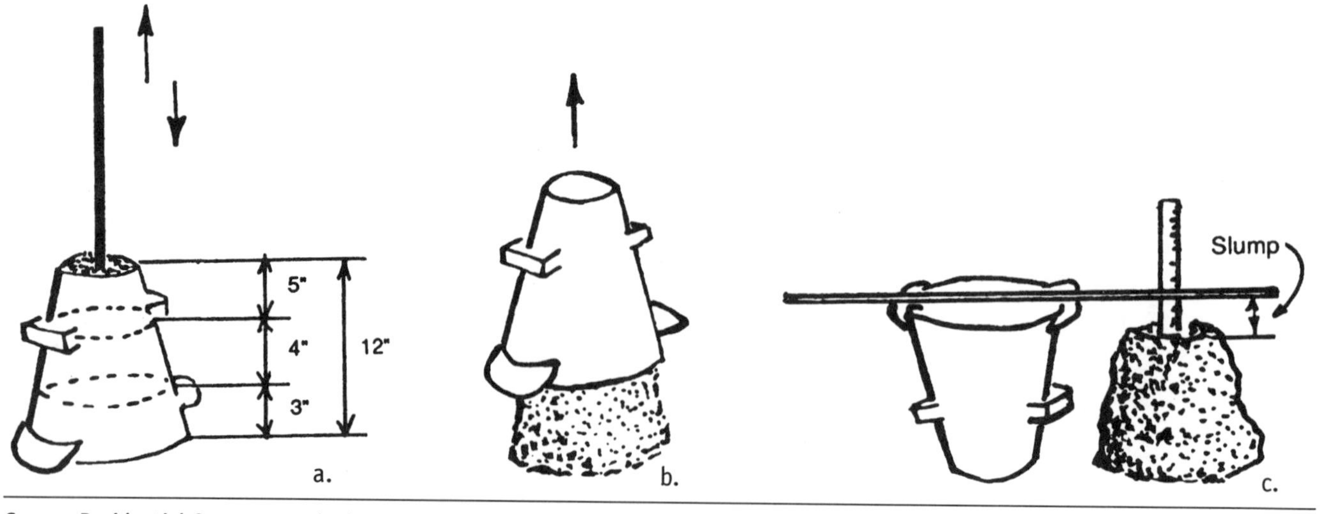

Source: Residential Concrete, *3rd ed., Home Builder Press, National Association of Home Builders, Washington, D.C.; 1999.*

tremendous pressure on the forms. Smooth, controlled placement helps minimize the likelihood of accidents or damage to the forms. Ideally, the concrete should be placed into the forms in 1- to 2-foot-deep layers referred to as *lifts*.

By the time the first lift is placed around the perimeter of the house the concrete usually has begun to harden or set up. At this point, workers can begin placing the second lift, which again should be 1 to 2 feet deep. As the concrete sets, the outward pressure it exerts on the form decreases.

During placement, do not drop the wet concrete more than a distance of 4 feet. A drop greater than this will cause the coarse and fine aggregates in the concrete to segregate, weakening the strength of the mix. If the concrete must fall farther than 4 feet, use a large-diameter hose to help prevent the aggregate in the concrete from separating from the paste. Maintaining proper slump also helps to avoid segregation of the concrete ingredients.

Keep the concrete trucks outside the perimeter of the residence so they do not disturb the trenches or forms. If it is impossible to position the transit truck so that the chute is placed directly into the forms, there are two systems that can move the concrete from the truck to the forms—a concrete pump and a conveyor system. When compared to using a wheelbarrow or trench cave-ins, these methods can be effective, though more costly, alternatives.

Moving Concrete with a Pump

A concrete pump or conveyor truck can resolve the problem of how to place the concrete in trenches and forms when the truck must stay a considerable distance away. The concrete from the transit truck is deposited into a hopper on the pumper, which usually is a separate unit. The concrete then is forced through a delivery hose. The hose can snake its way around obstacles such as trees, and up or down a hillside. The hose also can be lowered into the pier to control the free drop of the concrete.

Most pumpers can handle aggregates up to 1 1/2 inches in diameter in concrete with a 3-inch slump. Check with the pump operator, however, to make sure that the machinery does not require a special mix.

One worker usually can handle the delivery hose while a second person operates the pump and controls the rate of flow. The flow of concrete should not be stopped for more than a few minutes at a time. Further delay, especially if the hose is exposed to the hot sun, can cause the concrete to plug up the line. This also means that one truck should be ready to move into position as soon as the first is empty.

It is best to keep the flow going, even if it means pumping some concrete through the line and not using it. It is a lot easier to dispose of a small pile of concrete than to clear a plug in the line. When the placing of the concrete is nearly complete a plug usually is inserted into the delivery hose at the pump. The plug is forced through the hose by air or water pressure to clear most of the concrete from the hose. This reduces the final cleanup effort.

Using a Conveyor System

The second method of delivery is a conveyor system mounted on a transit truck. The concrete flows into a hopper, onto a conveyor belt, and then to a second conveyor on a moveable boom. The hydraulically operated boom can be rotated 360° and extended up to 100 feet. At the end of the boom the concrete is deposited into another hopper before finally flowing into a hose that can be inserted into the bottom of a pier so the concrete does not fall too far. Concrete being moved to a jobsite on a conveyor system is shown in Figure 5-4.

The conveyor system requires two workers. One controls the flow of concrete from the transit truck into the first hopper. The second controls both the positioning of the boom and the rate of flow. Trees can hamper the effectiveness of this delivery system if they are in the way of the boom. Before using this

FIGURE 5-4. A conveyor system is an effective way to move concrete to the site if obstacles reduce access.

Source: Putzmeister America Telebelt®, Sturtevant, WI.

system, check to make sure there is a clear path for the truck and the boom.

Consolidating the Concrete

Consolidating makes the lift look smoother and brings a thin layer of fine sand and cement particles to the surface. Commonly, consolidation is accomplished by mechanically vibrating the concrete. Vibrating each lift eliminates voids and air pockets. The vibrating consolidates the mix and makes the completed job look better. Over vibrating can cause the aggregates in the mixture to segregate and lessen the strength of the foundation wall.

To vibrate the concrete, a worker follows behind the crew placing the concrete and inserts a vibrator into the fresh concrete. The areas that the vibrator reaches should overlap to ensure that all the concrete is consolidated.

For the vibrator to work effectively it should be kept as vertical as possible and remain in the lift for 10 to 15 seconds. It should not be forced into the concrete, but be pulled into the lift by the force of gravity. Extend the vibrator through the lift being placed and about 6 inches into the previously placed concrete. This mixes the concrete together and creates a good bond between the lifts eliminating flow lines, seams, and planes of weakness. A vibrator is shown in Figure 5-5.

Do not vibrate lifts that already have begun to set. This causes the concrete to become fluid and increases the outward pressure that the concrete exerts on the forms and snap ties. In addition, vibrating concrete after it has begun to set can weaken the concrete and cause the aggregate to separate. When vibrating reinforced cast-in-place walls, make sure the vibrator does not disturb the placement of the reinforcing steel. A worker using a vibrator is shown in Figure 5-6.

Be careful not to over vibrate, however, because too much vibration can cause too many of the coarse particles to settle and create weaker layers of fine particles. Concrete placed with a slump of more than 5 inches should not be vibrated. Also do not use the vibrator to move the concrete. This over-vibrates the concrete and causes it to separate.

Vibration allows the use of concrete with a lower slump. The vibration loosens the particles and lets them flow easier. When using concrete with a lower slump, vibrate just after the concrete is placed. This consolidates the mix before it begins to set. Dryer concrete will set faster, lessening the outward forces exerted on the forms.

Installing Sill Anchoring Devices

Once the concrete reaches the desired height, place the *anchor bolts* or other hardware needed to attach the superstructure above. Place these anchors according to local codes, usually every 6 feet and within 12 inches of each corner and embedded into 7 inches of concrete.

FIGURE 5-5. The vibrator is used to consolidate lifts to eliminate flow lines, seams, and places of weakness.

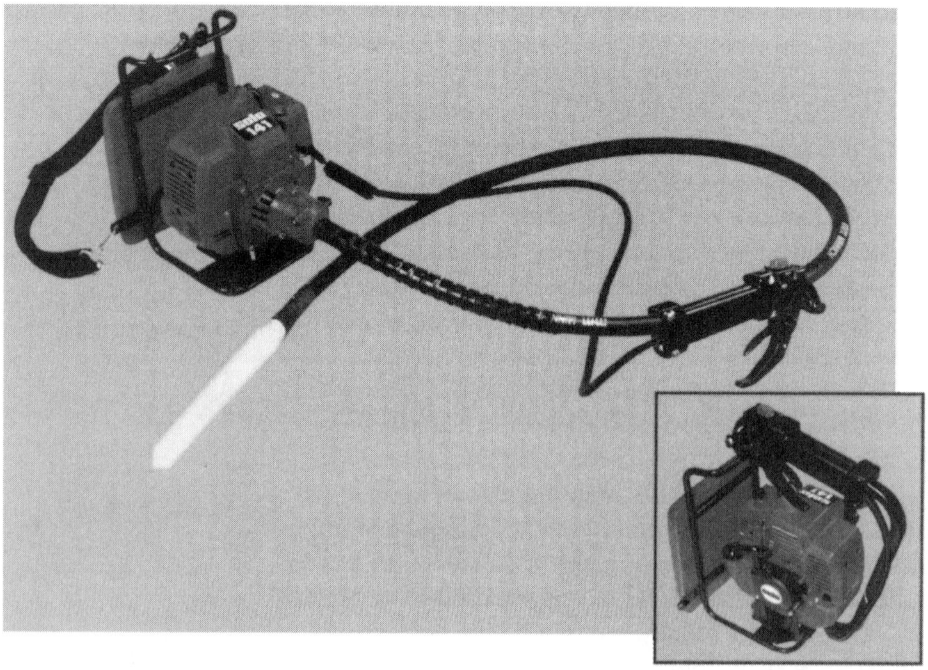

Source: Minnich Manufacturing, Mansfield, OH.

FIGURE 5-6. The vibrator is extended into the lower lift to create a good bond between the two lifts.

Source: The Wyco Tool Company, Racine, WI.

Anchor bolts traditionally are used to anchor sill plates to foundation walls. They must be placed plumb with at least 2 inches of the bolt extending above the wall. Anchor bolts can give framers some problems if their location obstructs the placement of floor joists, or if the nut and washer get lost. Anchors with perforated straps that wrap around the sill plate overcome some of the disadvantages of anchor bolts. Various anchor devices are shown in Figure 5-7.

Some municipalities allow the use of powder-driven fasteners. These nails are made from hardened steel and have excellent holding power when driven through the sill plate into the concrete by the

FIGURE 5-7. Anchor devices are used to attach sill plates to the foundation walls.

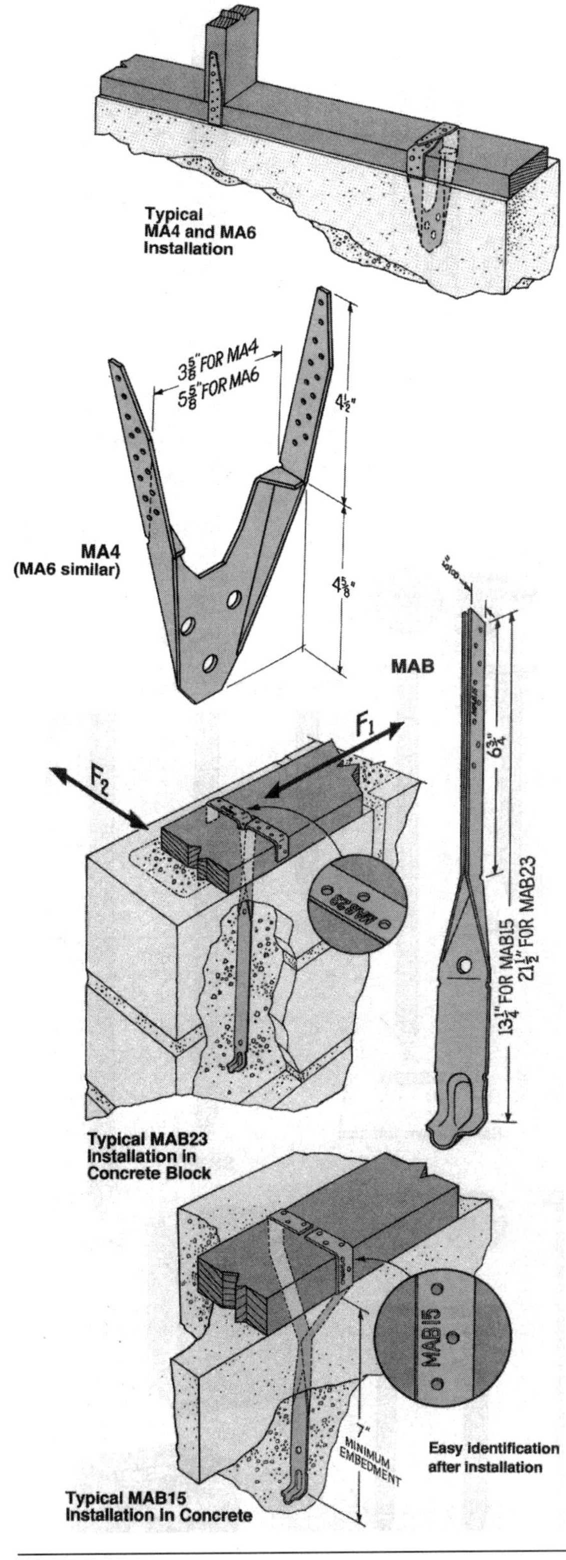

Source: Simpson Strong-Tie Co., Pleasanton, CA.

explosion of gunpowder in a specially designed tool. To meet codes the fasteners also must be attached through a large washer to hold the sill securely in place. Powder-driven fasteners cost slightly more than other sill-anchoring devices, but are easier to install and fewer may be required. A certified operator must operate the powder-driven tool.

For more detailed information on specifying and placing concrete on residential jobs, see *Residential Concrete, 3rd Edition* (1999), available from Home Builder Press.

Chapter Checklist

When considering a concrete footing or foundation, ask these questions regarding the mix that will be used in the footing and foundation.

- Is the rate of the pour acceptable?
- What is the slump of the concrete and is it free of or does it contain appropriate admixtures? Water should not be added to concrete. Adding more water than what is specified in the design can lessen the strength of the concrete.
- What's the air temperature? Are blankets available to protect the freshly cast walls from freezing?
- Are the correct numbers of specified sill plate anchors at the site?
- Is the crew adequate for the concrete casting, considering all contingencies?
- Is a proper work platform available? Standing on the top of the forms does not conform to Occupational Safety and Health Administration (OSHA) standards.
- Where can excess concrete be placed?
- Will a conveyor or pump be required to place the concrete into the forms from a drop of no more than 4 feet?
- Is a vibrator at the jobsite? Is there a worker knowledgeable in its use?
- Are proper safety precautions in place so that the ready-mix truck and the chute can be operated safely?

6

Using Footings in Foundations

A footing spreads the load of the structure over the soil by balancing the load of the building with the bearing capacity of the soil. Proper footing design and placement ensures maximum support for the foundation walls and the structure above. Footings can be constructed from concrete, gravel, or solid masonry. In some foundation systems, such as monolithic slabs, the footing is integrated into the floor slab or foundation wall.

There are specific requirements for the placement of footings. For example, footings must be placed on undisturbed soil unless the footing has been engineered for a site with disturbed soil. Soil is *disturbed* if it has been dug previously or back filled. Once disturbed, soil will not properly support loads imposed on it until it is properly compacted. Footings should not be placed on frozen ground because frozen ground settles when it thaws. If the ground settles unevenly it may not provide uniform support for the footing. For similar reasons footings should not be placed in standing water.

Most building codes have specific sizing requirements for footings. The table in Figure 6-1 provides calculations for the required width of masonry footings for one-, two-, and three-story homes.

These tables are based on soils with bearing capacities from 1,500 to 4,000 pounds per square foot (psf). To use the tables several other assumptions must be made about the soil conditions, including:

- Soil is well compacted.
- No fill has been used at the site.
- No expansive soil is present.
- Site is not conducive to significant erosion.

FIGURE 6-1. Most building codes specify minimum requirements for the width of footings.

	LOAD-BEARING VALUE OF SOIL (psf)					
	1,500	2,000	2,500	3,000	3,500	4,000
Convention Wood Frame Construction						
1-story	16	12	10	8	7	6
2-story	19	15	12	10	8	7
3-story	22	17	14	11	10	9
4-Inch Brick Veneer over Wood Frame or 8-Inch Hollow Concrete Masonry						
1-story	19	15	12	10	8	7
2-story	25	19	15	13	11	10
3-story	31	23	19	16	13	12
8-Inch Solid or Fully Grouted Masonry						
1-story	22	17	13	11	10	9
2-story	31	23	19	16	13	12
3-story	40	30	24	20	17	15

For SI: 1 inch = 25.4 mm, 1 psf = 0.0479 kN/m^2.

Source: Council of American Building Officials (95) Table 403.1.

If any of these assumptions cannot be made, an engineer should design the footing to accommodate the soil conditions. *Spread* footings, for example, often are used to distribute the load of the residence over a larger area of weaker soil, similar to the way snowshoes spread a person's weight across the surface of the snow. Spread footings may not be required by some code jurisdictions if the bearing capacity of the soil is enough to allow the foundation wall to bear directly on the soil.

Sizing a Footing

The exact dimensions of a footing are determined by comparing the design weight of the finished structure with the bearing capacity of the soil. As a rule of thumb, a spread footing is twice as wide as the thickness of the foundation wall it supports and as deep as the wall is thick. The bottom of the footing should be at least 6 inches below the frost line established by the local building code or, in areas unaffected by freezing temperatures in winter, at least 12 inches below the finished grade. A section of a footing is illustrated in Figure 6-2.

The bottom of the footing should be as level as possible and never should exceed a slope greater than 1 vertical foot per 10 horizontal feet or a 10-percent slope. If the ground is not excavated to a single, level elevation, a stepped footing is used. If the bottom of the footing is overexcavated, the additional depth or width created should be filled

FIGURE 6-2. These footing sections are illustrated with the finished grade.

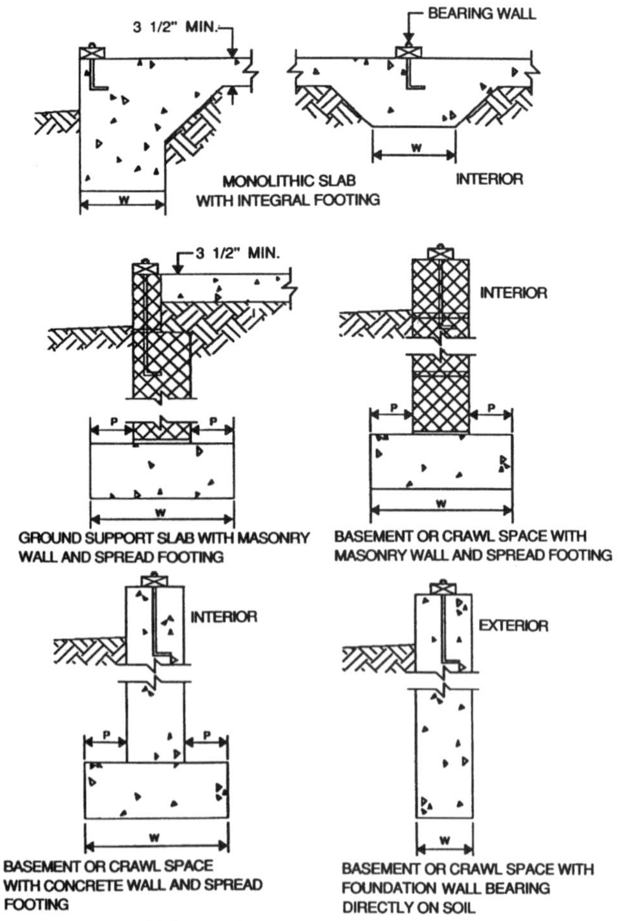

For SI: 1 inch = 25.4 mm, 1 foot = 304.8 mm.

Source: Council of American Building Officials (95) Figure 403.1a.

with concrete when the footing is placed. The top of the footing should be level to facilitate the erection of the foundation. Whether formed or cast directly into a trench dug into the earth, the footing should be positioned centrally below the foundation wall it will support.

Casting a footing that is wider than needed is not an acceptable practice. The footing should not be more than 6 inches wider than the foundation wall. Wider footings can fail by cracking parallel to the foundation. This is especially true for footings under 6-inch-thick foundation walls.

Most footings for residential buildings are cast in place using ready-mix concrete with a minimum compressive strength of 2,500 pounds per square inch (psi). If part of the footing will be above the finished grade and the structure is in a region where the footing will be exposed to freeze-thaw cycles, 3,000-psi air-entrained concrete should be used.

As the bearing capacity of the soil increases, the width of the footing can be decreased. On some soils the width of the required footing might equal the width of the foundation wall. In this situation the footing can be eliminated. Reducing the width of the footing or eliminating it entirely can save a considerable amount of concrete. If a 10-inch-wide footing that is 100 feet long is reduced to a 6-inch-wide footing of the same length, the difference is a cubic yard of concrete.

This savings must be balanced against possible engineering costs to analyze the soil and correctly size the footing. Local code officials may require a footing design developed by a registered engineer using information obtained from site testing before the width of the footing can be reduced. In a development where several homes will be built on similar soils, this could be a wise investment.

Reducing the depth of footings also can save money. In areas where the frost line requires the base of the footing to be 16 inches below grade, some contractors dig a trench that deep and fill it with concrete. This results in a footing that is twice as deep as it needs to be because the required depth of the footing is only 8 inches. The other 8 inches are included in the height of the foundation wall, which is only half the width of the footing. If the footing is 100 feet long this can reduce the concrete used by more than 1 1/2 cubic yards. Figure 6-3 shows a site overexcavated for formed footing.

FIGURE 6-3. Overexcavating increases concrete costs and may result in the failure of the footing.

Digging Trenched Footings

Before digging a trenched footing, delineate the sides of the trench with a string line or spray paint. If using a trenching machine allow for the offset of the digging teeth when placing the string line or paint marks.

The Occupational Safety and Health Administration (OSHA) sets strict standards for excavations and trenches. Because deep excavations are uncommon in residential construction, some OSHA standards exclude most residential projects. Safety considerations, however, require that site conditions, such as soil type, surface water and groundwater, water table, overhead and underground utilities, traffic, and the nearness of other structures and their condition be considered whenever excavating a basement or footing trench. The risk of cave-ins and other excavation-related accidents can be prevented or greatly reduced through strict compliance with OSHA standards.

Either a backhoe or a trencher digs most footings. Place the removed dirt several feet from the trench on the exterior side of the residence. This prevents the dirt from accidentally falling back into the trench and decreases the possibility that the weight of the pile will cause the sides of the trench to collapse into the excavation. Figure 6-4 shows a photo of a trencher digging a footing.

Remove all loose dirt and any remaining topsoil not removed by the mechanical excavator before placing the concrete. If plumbing pipes will exit through or under the footing, place a *sleeve*, or a pipe larger in diameter than the planned sewer or water supply pipe, in the excavation at a point below the bottom of the trench. During placement, make sure the concrete flows around the chase and does not fill in the ends. When separate water and sewer lines are required, install separate sleeves.

FIGURE 6-4. A trencher is an efficient tool for digging a footing.

When required, rebar can be placed vertically in the trench and the tops driven to a uniform height to indicate the top of the footing. The best way to ensure a uniform height is to use a builder's or laser level. The tops of the rebar provide a visual reference for the placement of the concrete.

If horizontal steel reinforcement is required by local codes, lay the rebar on *chairs*, shown in Figure 10-9 in Chapter 10, or wire brackets designed to hold the reinforcing steel in place several inches above the bottom of the trench. The horizontal rebar also can be tied to vertical rebar that has been driven into the ground. Another option is the use of metal stakes with formed notches that hold the horizontal rebar in place. Overlap the ends of each piece of rebar 2 feet and tie the pieces together with wire. Rebar placed at a corner using notched stakes is shown in Figure 6-5.

Before ordering the concrete, call for the necessary inspections of the footing. If the residence is being built to qualify for a Federal Housing Administration (FHA) or Veterans Administration (VA) loan, these agencies will accept inspection by most municipal building inspection departments. If an insured warranty will be provided, inspections for these warranty programs may satisfy the FHA and VA. If these options are not available the FHA or VA must be called for inspections. To prevent delays, it is wise to schedule inspections several days in advance. Schedule the delivery of the concrete for after the inspection. Some municipalities require that a footing be removed and recast if it is not inspected.

Casting Formed Footings

Casting footings in forms has several advantages. Forms make the casting of level footings easier. The concrete can be struck off along the top of the form assuring a level surface. Forms also make it easier to install foundation drainage because the gravel bed and drainage pipe can be placed without any further excavation. In addition, if the forms are coated with releasing agents before the concrete is placed they can be used many times, which spreads their cost over several jobs.

Forms for footings may be constructed with dimension lumber, plywood, metal, or plastic.

Specific forming techniques depend on the material used. The overall process, however, involves the creation of a stable mold that will hold the cast concrete in place until it has gained sufficient strength and stability to maintain its shape after the forms are removed.

When footings are cast in forms, the excavation for a basement foundation must be deep enough to accommodate the footings. The depth of the excavation should equal the height of the foundation wall plus the footing, less the height above the finished grade of the foundation wall.

Depth = (Foundation Wall Height + Footing Height) – Height of Foundation Wall above Finished Grade

FIGURE 6-5. Notched stakes are an effective way to turn rebar at corners.

Source: ReBar Stakes, Louisville, KY.

The top of the foundation wall should be at least 8 inches above the finished grade. If a masonry veneer is used, the top of the foundation only needs to be 4 inches above grade. In both situations, builders should make sure the height above grade meets applicable building codes.

Protecting the Forms

The most common forming material is dimensional lumber. If wooden forms are coated with releasing agents before the concrete is placed, they can last for a significant number of jobs. Steel forms represent a considerable investment, but will last much longer and may be worth the investment if they will be used to build a large number of homes. To keep steel forms in top condition they, too, must be coated with a releasing agent prior to each use. In addition to making the forms last longer, releasing agents also make the removal of the forms easier. Releasing agents can be applied with a hand-pump sprayer or brush. Do not use motor oil as a releasing agent. Used motor oil can leach into and contaminate groundwater on the site.

Constructing the Forms

Drive wooden bracing stakes, generally 1×2s or 2×2s or steel stakes, either square, round, or flat into the ground on the outside of the wooden forms every 3 feet to provide enough support to hold the forms steady when the concrete is placed. Attach the wooden forms to the stakes using duplex, or double-headed nails driven through wooden stakes or through holes drilled through the steel stakes. Steel forms have slots along one side through which steel stakes are driven and then cinched up with wedges. Once the exterior wall is placed, build the form for the inside of the footing using the same procedure. Hold the inside forms in place with stakes placed every 4 to 6 feet.

In addition to the stakes, attach strips to the top edge of the forms to keep the sides of the forms from spreading apart when the concrete is placed. Place these plastic, metal, or wooden *spreaders* every

FIGURE 6-6. Forms, stakes, and spreaders are the elements required to construct a formed foundation.

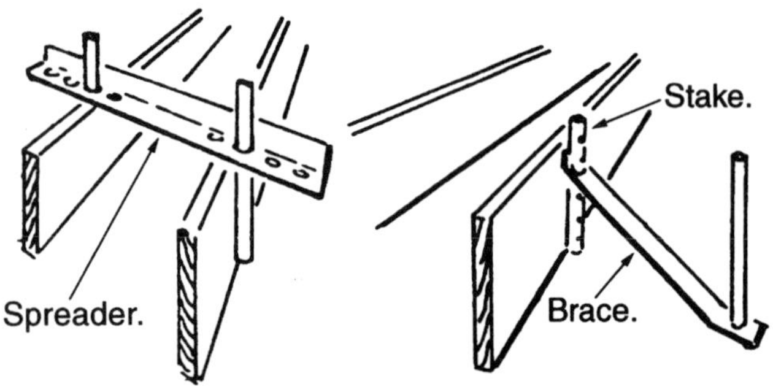

Source: Residential Concrete, *3rd ed., Home Builder Press, National Association of Home Builders, Washington, D.C.; 1999.*

4 to 6 feet. Use an optical level to check the height of the forms because the more level the footing, the easier it will be to erect the remaining segments of the foundation system. Figure 6-6 shows the forms, stakes, and spreaders used to construct a formed foundation.

A reusable plastic spreader, called a *Brack-It,* does not obstruct the finishing of the top of the footing. It expands from 4 to 30 inches and makes an arch between the tops of the footing forms instead of laying flat. The nonstick surface of the Brack-It can be cleaned easily for reuse.

If the forms are taller than needed, mark a level line on the interior of the form to indicate the top of the footing. The level line can be determined using a builder's or laser level and then marked, either with a chalk line or nails driven into the forms at the desired level at 4- to 6-foot intervals. Many builders find that the nails are more visible when the concrete is being placed.

Aligning Form Walls

Many builders use a string line to align the forms. Place a straight string line about 1/4 inch away from the form to be straightened. Then check that there is a uniform 1/4-inch gap between the form and the string. If the form gets too close or touches the string, push the form away from the string. If the gap widens more than 1/4 inch, push the form towards the string to narrow the gap. This method is more practical than aligning the form with the string line touching it.

Curing the Concrete

Curing requires favorable moisture and temperature conditions. In general, this means a period of about seven days for the most common Type I cement concrete at or above 50° F. The concrete contains adequate moisture for curing. Proper curing offers many benefits including improved strength, reduced cracking, and reduced permeability, all of which can be great assets for basement walls, slabs, and footings.

After the concrete has hardened sufficiently, usually after 24 to 48 hours, remove and clean the forms. Leaving the forms in place or spraying a curing compound on the surface of the concrete immediately after the forms are removed helps prevent rapid moisture loss and improper curing.

Using Alternate Footing Forms

An alternative form system uses a cloth membrane and a wood and metal frame to form footings. Drive metal stakes into the ground following the footprint of the footing. Then install the metal yokes every 6 feet. These support a set of horizontal 2×4s that are placed flat into the yokes, one for each side of the footing. The yokes are adjustable so one set can be used for a wide range of footing widths. The clamps or brackets allow the 2×4s to be adjusted up or down to level the proposed top of the footing.

The cloth membrane, which is woven polyethylene fabric, is available in 100-foot-long-by-52-inch-wide rolls. The membrane is draped between the 2×4s and stapled to the horizontal boards. The membrane has guidelines marked on it that can be followed to keep the fabric aligned. A nail driven through the cloth about every 5 feet keeps the membrane from moving. The concrete is placed into the channel made by the cloth.

Because the cloth membrane is not rigid it can conform to an irregular surface, which should reduce the time required to set up the forms. A three-person crew can set up the yokes, place the wooden members, and staple the fabric for approximately 80 feet of footing in an hour. Care should be taken however to assure that the minimum width of the footing is maintained. After the concrete has set the 2×4s and the clamps are removed and the membrane is left in place.

Constructing Stepped Concrete Footings

Most codes limit the elevation change at the bottom of a footing to a maximum slope of 1 in 10. When building on ground with a steeper slope or when the construction of a level footing is impractical, a *stepped* footing often is used. Stepped footings also provide solutions for homes on sloped lots that do not have basements, split-level homes, and homes with attached garages. For a steeply sloped lot, several steps may need to be incorporated into the footing.

Stepped footings are installed with one or more changes in level depending on the slope. Most codes require that the difference in the height between any two steps not be greater than three-quarters of the length of the upper footing. The footing must be continuous with vertical sections connecting the two levels. The width of these vertical sections should be the same as the width of the horizontal footings. The thickness of the vertical portion should be the same as the depth of the horizontal portion.

The vertical sections of a stepped footing are formed by erecting a bulkhead across the end of the upper trench or form. The bulkhead must be adequately braced to hold the concrete in place. Steel reinforcement, if required to meet codes or site conditions, should run continuously between the two levels of the footing.

The steps in the foundation layout are positioned according to the construction materials used. For example, if concrete blocks or insulating concrete forms (ICFs) are used, measure the distance from the foundation corners to the footing steps in terms of block lengths. Plan the height difference in the steps to correspond to the height of the product. If the solid concrete foundation walls will be cast, consider the dimensions of the forms when placing the steps. If plywood is used as a forming material, try to set the horizontal distance between the steps in 4-foot increments. If manufactured forms are used it might be best to consult with the form supplier to see which layout would be most efficient. The finish grade of the project also must be factored into the placement of the steps, because the bottom of each step must be positioned so that it rests 6 inches below the frost line established by the finished grade.

Use concrete with a maximum slump of 3 inches when casting stepped footings. Because this concrete is relatively dry, it does not tend to flow as readily as concrete with a slump greater than 5 inches. For best results first place the concrete in the steps and then fill in the lower horizontal surface. Continue placing the concrete in the progressively higher levels.

Do not mechanically vibrate concrete in stepped footings. This will make even low-slump concrete fluid and increase the likelihood that it will flow out of the step. Instead *rod* the concrete to reduce the occurrence of honeycombing. Rodding consolidates the concrete by probing it with a wooden or metal rod several times. If the pour has to be stopped to move one ready-mix truck out and position another the concrete can be *rodded* to blend the two layers together. This reduces the likelihood that a cold joint will form.

A cold joint probably will occur if the casting must be stopped for more than 30 minutes, or even less depending on weather. If this occurs, form a keyway into the initial layer to tie the two layers together.

Building Isolated or Spot Footings

Whenever a significant load is imposed at a point other than along the perimeter of the residence an isolated or spot footing must be placed. Spot footings are required under columns or posts that are used to support beams, and under other concentrated loads such as masonry fireplaces. The size of the footing is based on the bearing capacity of the soil and the imposed load. The number of stories and the spacing of the posts that support the girders in the floor framing system also are considered. Footing sizes for girder support columns are provided in Figure 6-7.

Column footings typically are about 2 feet square by 1 foot deep and centered on the base of the column. The top of the spot footing should be 4 inches below the top of the basement's finished floor. This allows for the placement of a 4-inch-thick continuous slab over the spot footing. Placing the basement slab after the posts and girders are installed securely anchors the posts in position. Construct isolation joints near the base of the posts to control cracking in the basement slab. Isolated footings

FIGURE 6-7. Footing sizes for girder support columns are based on standardized calculations.

		ONE STORY		TWO STORY		THREE STORY	
WIDTH OF STRUCTURE (feet)	GIRDER SIZE (inches)	Maximum Span (feet-inches)	Footing Size (inches)	Maximum Span (feet-inches)	Footing Size (inches)	Maximum Span (feet-inches)	Footing Size (inches)
24	3-2 × 8	6-7	17 × 17*	4-11	20 × 20	4-1	22 × 22
	4-2 × 8	7-8	19 × 19*	5-8	21 × 21	4-9	24 × 24
	3-2 × 10	8-5	20 × 20*	6-3	23 × 23	5-3	25 × 25
	4-2 × 10	9-9	21 × 21	7-3	24 × 24	6-1	27 × 27
	3-2 × 12	10-3	22 × 22	7-8	25 × 25	6-4	27 × 27
	4-2 × 12	11-10	23 × 23	8-10	27 × 27	7-4	29 × 29
26	3-2 × 8	6-4	17 × 17*	4-9	20 × 20	3-11	22 × 22
	4-2 × 8	7-4	18 × 18*	5-6	22 × 22	4-7	24 × 24
	3-2 × 10	8-1	19 × 19	6-1	23 × 23	5-0	25 × 25
	4-2 × 10	9-4	21 × 21	7-0	24 × 24	5-10	27 × 27
	3-2 × 12	9-10	21 × 21	7-4	25 × 25	6-1	28 × 28
	4-2 × 12	11-5	23 × 23	8-6	27 × 27	7-1	30 × 30
28	3-2 × 8	6-2	17 × 17*	4-7	21 × 21	3-10	23 × 23
	4-2 × 8	7-1	18 × 18*	5-3	22 × 22	4-5	24 × 24
	3-2 × 10	7-10	19 × 19	5-10	23 × 23	4-10	26 × 26
	4-2 × 10	9-0	20 × 20	6-9	25 × 25	5-7	28 × 28
	3-2 × 12	9-6	21 × 21	7-1	26 × 26	5-11	28 × 28
	4-2 × 12	11-0	22 × 22	8-2	28 × 28	6-10	30 × 30
32	3-2 × 8	5-9	16 × 16*	4-3	21 × 21	3-7	24 × 24
	4-2 × 8	6-7	17 × 17	4-11	23 × 23	4-1	25 × 25
	3-2 × 10	7-4	18 × 18	5-5	24 × 24	4-6	27 × 27
	4-2 × 10	8-5	20 × 20	6-3	26 × 26	5-3	28 × 28
	3-2 × 12	8-11	20 × 20	6-8	27 × 27	5-6	29 × 29
	4-2 × 12	10-3	22 × 22	7-8	29 × 29	6-4	31 × 31

For SI: 1 inch = 25.4 mm, 1 foot = 304.8 mm, 1 psf = 0.0479 kN/m^2, 1 psi = 6.895 kPa.

Source: Council of American Building Officials (95) Table 502.3.3b.

under fireplaces typically extend 12 inches beyond the fireplace and are 12 inches thick.

Columns placed on spot footings to support beams should be corrosion resistant steel pipes. Many are adjustable to various heights, which can compensate for slight changes in elevation from footing to footing. The base of the column usually is cast into the concrete slab of a basement foundation.

Concrete block piers also can be placed on an isolated footing to serve as columns. They are erected as the walls are placed. Temporary bracing is advisable for block piers or columns that are more than 4 feet tall because they can be unstable until the beam is set on them. Use decay-resistant or treated wood columns and metal bases to hold the bottom of the post above the surface of the floor. This keeps the base of the wooden member dry and enables the post to be replaced if necessary.

The fabric form system uses bags filled with concrete for spot footings. The bags are available in sizes that correspond to the common sizes of spot footings. A post, spot footing, and isolation joint are illustrated in Figure 6-8.

Using Footing Blocks

During the construction in Europe that followed World War II, concrete masonry footing blocks were developed as a quick and economical way to produce prefabricated footings. A derivative of this system of interlocking solid cast concrete blocks is available in the United States.

In this system, 4×16-inch interlocking stretcher units are used to support 8-inch block walls and 3×20-inch stretcher units are used to support 10- and 12-inch walls. Corner units interlock with the stretchers to form the corners. All units are solid 8-inch-tall blocks. The footing units are laid under the same conditions as traditional footings. The site, as with all footings, must be level and free of organic soils, and the ground may not be frozen. The base of the footing blocks also must rest 6 inches below the

FIGURE 6-8. Isolation joints control cracks.

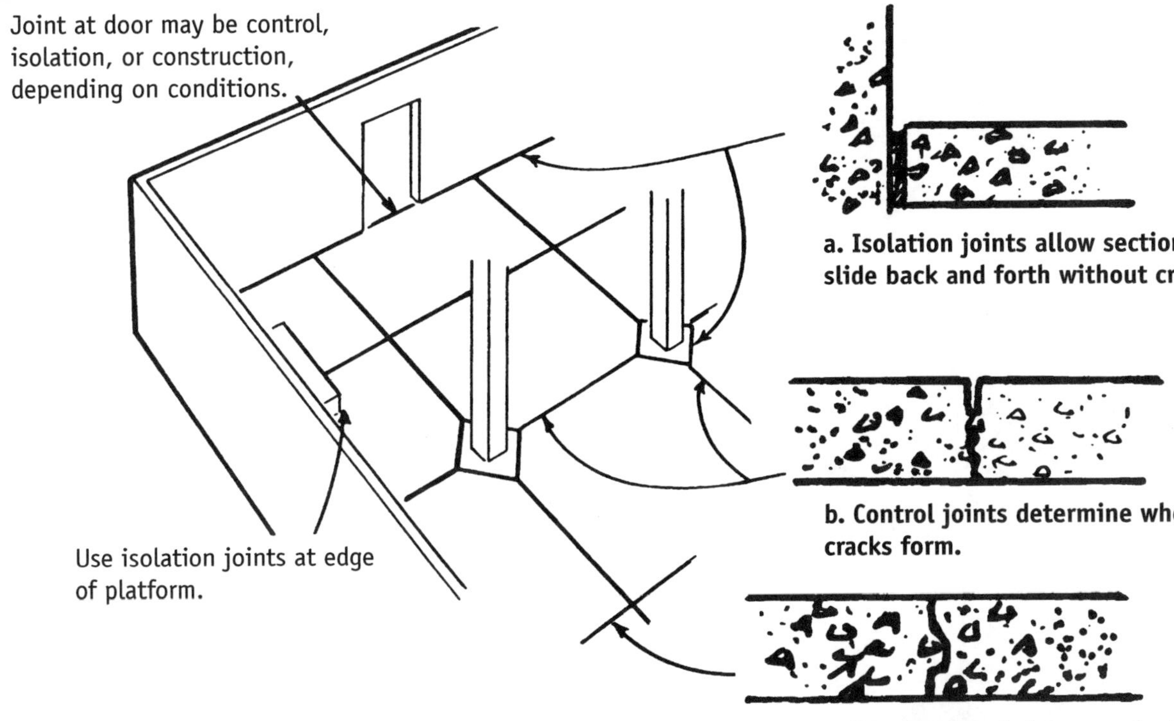

a. Isolation joints allow sections to slide back and forth without cracking.

b. Control joints determine where cracks form.

c. Construction joints, required when intervals occur between castings, sould coincide with control joint locations if possible.

Source: Residential Concrete, *3rd ed., Home Builder Press, National Association of Home Builders, Washington, D.C.; 1999.*

frost line. This may require excavation of a trench wide enough to allow workers to place the first course of blocks.

While mortar, sand, or pea gravel can be used to fill in under the blocks to compensate for an uneven base, mortar is not needed to bond the units together. Place the blocks as level as possible, but up to a 3/4-inch variance from level will not compromise the strength of the system. Once the footing blocks are in place, cover the tops of the blocks with a coating of Type M or Type S mortar that is as thick as the width of the wall.

Truss and ladder wire can be placed into the mortar layer to increase the lateral strength of the footing. Cover the reinforcement wire with a minimum 1/2-inch-thick coat of mortar. Place any additional reinforcing steel needed to meet local building codes within the concrete masonry *beam* or *trough* blocks laid as the first course of the foundation with the remaining courses built on top. A detailed drawing of a footing block is provided in Figure 6-9.

This system saves time and costs because the masonry contractor who erects the foundation wall usually can handle the entire job. A separate contractor is not needed for the footing. Another time saving feature is that construction of the foundation typically can begin the same day that the footing is placed. The simplified ordering and handling of materials also helps maintain schedules and avoids construction waste. As with all foundation systems that use concrete or mortar, take precautions to prevent the materials from freezing.

Laying Gravel Footings

Some foundation systems, such as preserved wood or precast concrete, require a footing that has very good drainage. These foundations can rest on gravel

FIGURE 6-9. This detailed drawing illustrates the footing blocks required to turn a corner.

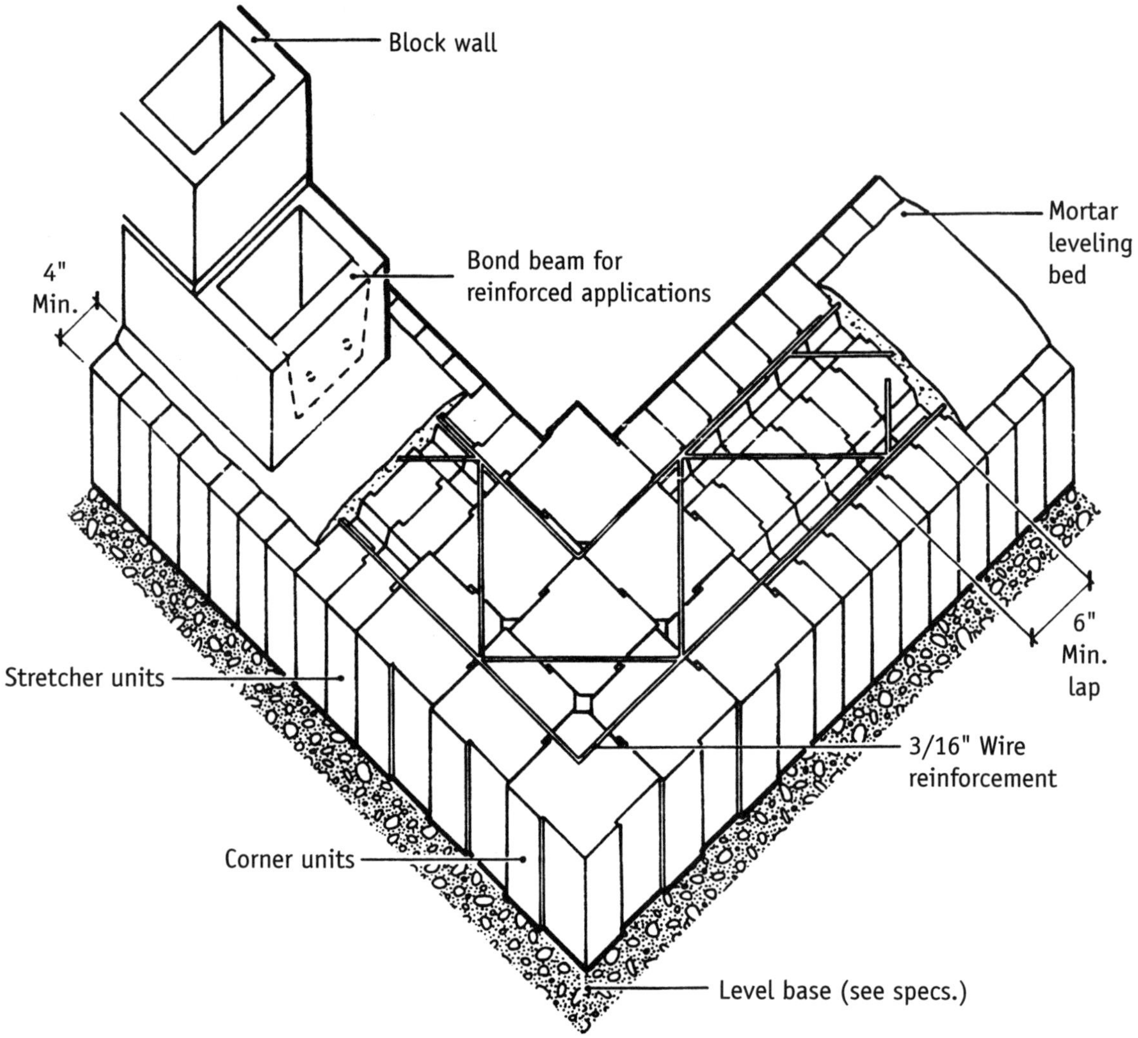

Source: Footer Block™ illustration courtesy of Rapid Building Systems, Division of Synthesis Inc., Reston, VA.

footings. Usually the footings are twice as wide and three-quarters as deep as the width of the footing plates, which are the bottom members of the foundation wall.

After digging the trench, place washed gravel in the trench. The gravel may be up to 1 1/2 inches in diameter and should be free of dust and organic material that can choke the drainage system. Crushed limestone also may be used. The limestone will support heavier loads, but must be washed well to remove its dust. Washed river gravel, if available, is another acceptable material.

Once the gravel is placed, compact it. A gasoline-powered tamper works best, but hand tamping with a 12-inch-square steel tamping rod is sufficient. Use a builder's or laser level or level string lines and a story pole to make sure the surface of the footing is level.

Chapter Checklist

When constructing any type of footing, ask the following questions.

- Has the building permit been issued?
- How deep is the frost line?
- Can the footing be reduced in size or eliminated due to the bearing capacity of the soil?
- Will the home's placement on the site allow adequate drainage or is a drainage system required?

- Can an effective drainage system be placed easily around the planned footing system?
- How difficult will it be for a trenching machine to work at the site?
- Would it be easier to form the footing?
- Is the lot level? Will the footing need to be stepped?
- If the footing is stepped, do the differences in the elevations correspond with the materials used for the foundation walls? For example, when concrete blocks are used are elevations measured in 8-inch increments?
- Is the bottom of the footing level with a maximum slope no more than 1 in 10?
- Has the footing been overexcavated? If so, what quantity of concrete or gravel is needed to fill in the excess area?
- Is the footing being cast at the correct depth or is concrete being wasted?
- Are sleeves placed under the footing for plumbing piping?
- Is the top of the footing level?
- Are all the spot footings in place?
- If footing drains will be used, can they be correctly placed?
- When building walkout basements, are all points 6 inches below the frost line?
- If erecting a block foundation, is there an advantage to using footing blocks?

In addition be sure that:

- The bottom of the footing is level.
- The ground is not frozen.
- The site where the footing is to be placed does not have standing water.

7

Placing Cast-in-Place Concrete Foundations

A cast-in-place concrete foundation offers many advantages when building in areas where the frost line is 3 feet or deeper, the terrain slopes, or when extending a foundation for a residential addition. Finishing the basement creates added living space for little additional cost. If the height of the backfill or the soil requires that the concrete be reinforced, reinforcing steel can be placed easily in cast-in-place concrete foundation walls.

Determining the Thickness of Walls

Model and local building codes coordinate the thickness of cast-in-place foundations and the planned depth of the backfill surrounding the basement. Most building codes allow a 6-inch-thick basement wall of unreinforced concrete back filled to a maximum depth of 4 feet. Plain concrete foundations with a thickness of 8, 10, or 12 inches can be back filled to a maximum depth of 7 feet.

With reinforcing steel, a cast-in-place concrete wall can be engineered to resist even greater backfill pressures and still adequately support the structure. In areas where concrete costs are high it may be more cost effective to use a specially designed 6-inch reinforced wall in place of a thicker wall that is not reinforced. Codes allow room for error when casting foundation walls. For example, walls that are 8 inches or wider can be 1/2 inch less than the nominal size. The loss of 1/2 inch for a 6-inch wall, however, is not acceptable.

Compensating for Exterior Pressure

The weight of the residence pushes down on the top of the foundation while the earth holds up the bottom. This *compressive load* squeezes the foundation wall from top to bottom while the surrounding soil, and the moisture in that soil, push against the sides of the walls. This side pressure is called the *lateral* or *side load*. Unlike the balancing pressure created between the house and the earth, there is nothing inside the basement to offset the lateral load created by the backfill. The foundation therefore must be designed to transfer this load to the footing and the first floor of the residence.

Foundation walls are designed to resist lateral loads by essentially acting as a beam that works against flexural or bending forces. Because concrete that is not reinforced has low flexural and tensile strengths, reinforcing steel is incorporated into walls when conditions place these stresses on the foundation.

Water in the soil exerts a force called *hydrostatic* pressure on the foundation. Both the lateral and hydrostatic loads and the height of the backfill affect the design of the foundation. In well-drained, stable soils where the lateral and hydrostatic loads are minimal, the height of the backfill is used to determine the thickness of the walls. If the potential for a significant amount of water exists in the backfill, a structural engineer must design the foundation walls. Consult with local building officials or with a structural engineer if there are questions about whether or not the soil at a particular building site is conducive to the construction of a certain type of foundation.

Lateral forces also affect the scheduling of construction activities. Backfill should not be placed until the basement slab is placed and the sheathing applied to the first floor joists, or the foundation walls are braced adequately with a vertical 2×6 every 8 to 10 feet backed with a diagonal brace secured to the basement floor.

Supporting the Height of the Residence

The number of stories in the house also must be considered when determining the thickness of cast-in-place foundation walls. Six-inch-thick concrete walls that are not reinforced may be adequate for a single-story residence when the backfill is less than 6 feet. Walls that are 8 inches thick are acceptable for two-story dwellings when the backfill is less than 7 feet deep. For three-story buildings, 10-inch-thick walls are permitted if the backfill is less than 8 feet deep.

If a masonry veneer is supported by the foundation, however, 6-inch thick walls are not an acceptable thickness. One rule of thumb is to increase the required thickness of the foundation by 2 inches whenever a masonry veneer is used. The thickness of the walls can be mixed. For example, 10-inch walls can be used under masonry veneers and 8-inch walls can be used when siding is installed for the exterior finish.

Determining the Height of Foundation Walls

The height of cast-in-place foundation walls is calculated using two main factors—the depth of the frost line and the inclusion of a basement. The base of the foundation or footing must be below the frost line and the top of the foundation should be at least 8 inches above the finished grade. The footing usually is as deep as the thickness of the foundation wall and twice as wide. The distance from the top of the footing to the point 8 inches above the finished grade determines the height of the foundation wall.

When a crawl-space system is used, the height of the foundation wall should provide at least an 18-inch clearance between the bottom of the floor joists and the grade under the floor to accommodate mechanical systems. If the cast-in-place foundation wall is part of a basement system with habitable space the height of the wall should be a minimum of 7 feet 10 inches to provide a habitable space of 7 feet 6 inches. This allows the placement of a 4-inch-thick cast-in-place concrete slab in the basement to meet minimum code requirements. If the basement will not be used for habitable space, the height of the foundation wall should be tall enough to provide a clearance of 6 feet 4 inches from the basement floor to the underside of all ducts, beams, girders, and other obstructions. A 9-foot-tall foundation wall is an excellent option because it allows the basement to be finished in the future with a dropped ceiling to hide the mechanical systems.

Protecting Workers

Scaffolding brackets, especially made to hang from walers, are available and should be used to provide a safe work platform. Workers should not walk on the tops of the forms while concrete is being placed. In addition, workers six feet or more off the ground must have fall protection. Provide this by installing sturdy guardrails along the sides of the scaffolding or by providing a safety harness.

Selecting Materials

Site-built forms traditionally have been used to construct cast-in-place concrete walls. The forms for both sides of the wall should present a smooth surface that is free of gaps through which the wet concrete can leak. The most frequently used plywood size is 3/4 inch. Other sizes can be used, although plywood thinner than 3/4 inch can *deflect* or warp from the pressure exerted by the weight of the concrete and result in a wavy surface on the walls. Multiple layers of 1/4-inch plywood, however, can be used to create curves and yet be strong enough to resist unwanted deflection.

To provide a more attractive finish the plywood should be at least a BB grade or better. Specially designed plyform materials that are laminated to create a smooth surface also provide an excellent finish. Site-built forms are shown in Figure 7-1.

If forms will be used repeatedly, an investment in the slightly more expensive plyform materials might prove more economical over time. Concrete does not adhere to the surface of these forms as much as it does plywood. Site-built forms also can be constructed of 2-inch-thick extruded polystyrene. These forms are very lightweight, but require the use of specialized wall ties and can be damaged easily. These materials are described in more detail in Chapter 8.

Constructing the Walls

When forming walls, place the two facing sides of the form a consistent distance from each other in order to ensure that the finished wall is smooth, straight, and the specified thickness. Most contractors first position the exterior-facing form along a chalk line that marks the exterior of the basement. Offset the chalk line an amount equal to the thickness of the form. Nail 2×4s to the footing to help keep the base of the foundation wall true. The 2×4s can be considered a *waler*, but their main purpose is to align the exterior forms and help hold them in place until the interior-facing form can be attached with the wall ties, which remain embedded in the finished wall.

Walers, or reinforcing boards, support individual forms, help keep the walls straight, and help

FIGURE 7-1. To provide an attractive finish, use at least BB grade plywood or plyform materials.

Source: Symons Corporation, Des Plaines, IL.

prevent the forms from bulging under the pressure of the freshly cast concrete. They are placed on both the interior and exterior wall forms. If a single 2×4 serves as the waler, further support can be provided by clamping *vertical braces* or *strongbacks* to the wall ties at 6-foot intervals. If *double walers* or two 2×4s are used, the vertical bracing normally can be eliminated. Number 2 or better grade softwood lumber generally is used for the walers and braces. A diagram labeling form parts is shown in Figure 7-2.

Affixing Wall Ties

Snap wall ties get their name from the fact that the portion of the tie that protrudes from the finished wall can be snapped off easily when the forms are removed, presenting a smoother surface. Because metal wall ties can rust, some manufacturers produce wall ties made of fiberglass or plastic, which are recommended for use in soil with high moisture content.

When purchasing materials for forms, be sure to buy enough wall ties at the appropriate widths. Check the manufacturer's instructions about the vertical and horizontal spacing of the ties in order to sustain the pressure exerted by the weight of the concrete during casting and curing. As a rule of thumb, each wall tie can support 4 square feet of wall. Adhere to the manufacturer's specifications, however, when drilling the tie holes in the form's facing material. The holes should be uniform in depth and generally drilled in two rows about 1 foot from the long edge of the form.

Extend the ties far enough through the drilled holes to attach the horizontal walers. Hold the exposed end of each wall tie firmly against the waler with a reusable metal piece such as a *bracket clamp*, *cam*, or *wedge*. A bracket clamp has a leg that fits into the loop on the wall tie and a slot into which the waler is set. A cam has an offset lever and hook that, when turned, pushes against the waler. A wedge is a thin, triangular piece of metal that is driven into the loop of the wedge.

FIGURE 7-2. Bracing, cones, and snap ties are part of site-built forms.

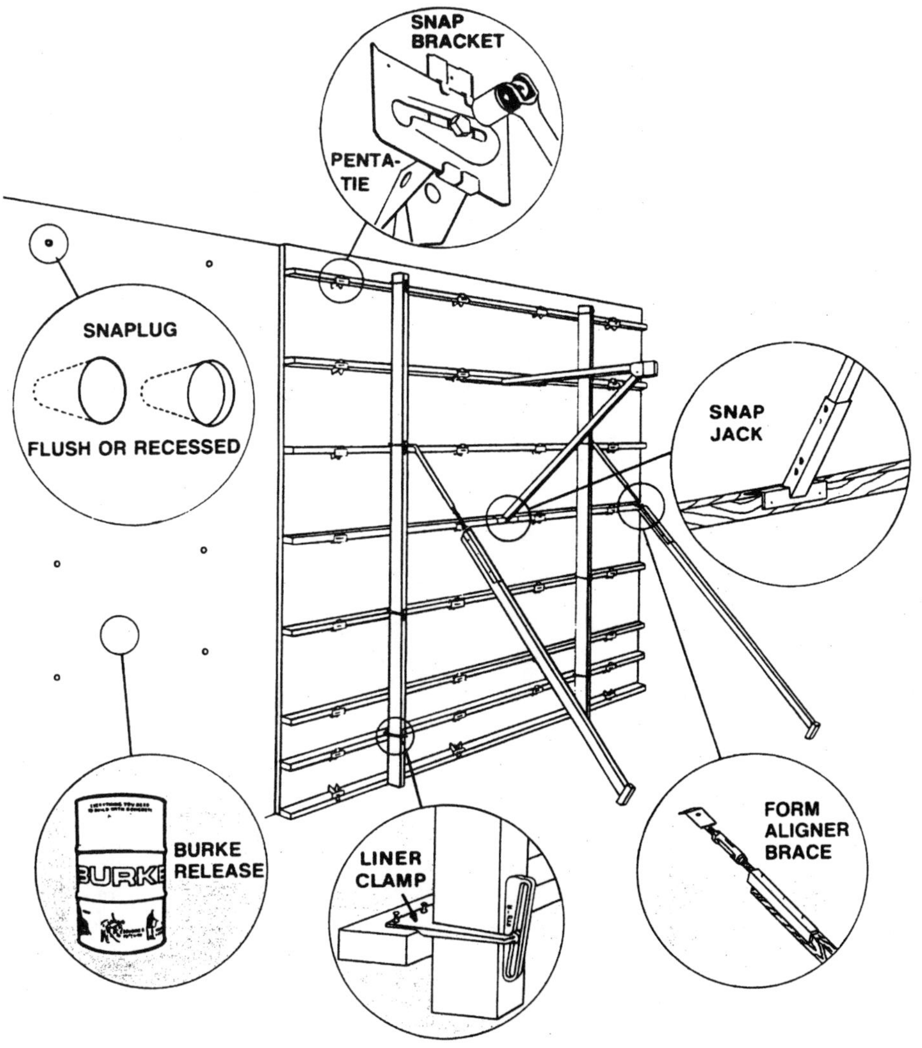

Source: Aluma Enterprises Inc., Toronto, Ontario, Canada.

Attaching Plastic Cones

A plastic cone is attached near both ends of most wall ties. The spacing of the cones defines the thickness of the wall. Ties with the cones spaced 6 inches apart are used for 6-inch walls, 8 inches apart for 8-inch walls, and so on. The cone is designed to be slightly wider than the diameter of the hole drilled through the form. This allows a set amount of the looped end of the wall tie to be inserted through the form. After each wall tie is secured to the waler, the cone, which is positioned on the inside of the form, helps the wall tie fit snugly against the form and maintains uniform spacing between the faces of the forms. The plastic cone also helps ensure adequate flexibility in the wall tie so that it can be snapped off when the form is removed.

Remove the cones when the looped ends of the ties are snapped off. This leaves a small indentation or dimple in the surface of the wall. Dimples on the exterior side of the wall generally are filled with a cementitious grout before a waterproofing compound is applied. Indentations on the interior side also can be filled but many builders leave them alone.

Installing Bracing

To keep the forms for the foundation walls straight while placing the concrete, brace the forms every 10 to 12 feet. *Rakers*, or diagonal braces, extend from the top of the form to the interior of the formed space and provide additional support to ensure that the wall is straight and true. It is virtually impossible to correct a wall that has been cast in crooked forms. The use of diagonal bracing helps ensure that walls are cast straight.

When installing rakers, check to make sure the forms are aligned accurately. To do this, establish a string line that is 1/4 of an inch out from the true exterior side of the wall. If the distance between the string and the forms is not uniform, either push or pull on the forms to properly align them and then secure the rakers to keep the forms in place. Two 8-foot 2×4s nailed together, one 6 feet from the end and the other at mid length, can be used as a lever to push against the form.

Building Blockouts and Brick Ledges

Construct blockouts for openings such as doors or windows as soon as the wall forms are in position. Rip 2× lumber lengthwise to the exact thickness of the finished wall. The interior dimensions of the blockout should correspond to the rough opening size of the window or door. If the blockouts will be left as nailers for the installation of doors and windows, use treated lumber. Check to make sure the blockout is correctly placed and plumb. Once the wall is placed, corrections are expensive. A blockout for a window is shown in Figure 7-3.

Attach the ripped board securely to the exterior form with duplex or double-headed nails. To prevent the sides of the blockout from bowing in or deflecting, place horizontal and vertical bracing at 2-foot intervals. Leave all form ties in place, including any that fall in the areas that are blocked out. To hold the blockout in the cast wall, drive nails partially into the blockouts from the side that will be against the concrete. The nails cast into the wall will anchor the blockout in place.

If a concrete header will be created above a door or window blockout, reinforce it with steel. Two # 4, 1/2-inch-diameter reinforcing bars should support the header over a 3-foot door or window. The rods should extend at least 12 inches past the edge of the blocked opening into the concrete foundation wall.

To create a pipe chase, secure a piece of polyvinylchloride (PVC) pipe at the appropriate location. The pipe used to create the chase should be larger than the pipe that will run through it.

If the home will have a masonry veneer, create a brick ledge along the top of the foundation wall. Form the ledge using 2×4s and plywood fixed to the interior face of the exterior side of the wall forms. Use chamfer strips for control joints. Some builders

FIGURE 7-3. Installing a blockout before placing the concrete creates a void that corresponds to the rough opening size of the window or door.

Source: Tuf-N-Lite, Springboro, OH.

FIGURE 7-4. A brick ledge incorporated into the top of the foundation wall supports a masonry veneer.

prefer to mark the top of the cast-in-place blockouts for brick ledges after both the interior and exterior walls are formed. Waiting until both walls are formed allows the work of creating the blockouts to proceed much faster. A brick ledge is shown in Figure 7-4.

Incorporating Control Joints

Because concrete tends to expand and contract with temperature changes, cracking can occur. To reduce this possibility, incorporate vertical *control joints* in walls that are 30 feet or longer. The vertical joints divide long walls into smaller segments and create deliberate weak spots that guide the formation of controlled cracks.

To create a vertical control joint, place *chamfer strips,* which are small pieces of wood cut in a "V" shape, on both the interior and exterior wall forms. Line up the strips so they are directly opposite each other. Each strip should leave an indentation equal to 1/8 the thickness of the wall. For example, chamfer strips for an 8-inch-thick wall would be 1 inch thick. Chamfer strips for a 12-inch wall would be 1 1/2 inches thick. Locate the control joints no more than 20 feet apart. The diagram in Figure 7-5 shows the proper placement of vertical control joints.

With a well-formed vertical control joint, any cracks that may form in the concrete as a result of normal weather- and temperature-related expansion and contraction should occur along the vertical joint line. After casting, fill the notch created by the chamfer strips with a high-grade caulk to seal any possible cracks and help protect the interior of the wall against moisture penetration.

Reinforcing Cast-in-Place Foundations

Some site conditions require that cast-in-place concrete walls be reinforced with steel. These conditions include unstable soils, high water tables, and situations where the height of the backfill around the foundation wall exceeds 6 feet for a 6-inch-thick wall, 7 feet for an 8-inch-thick wall, or 8 feet for a wall 10 or 12 inches thick. Building sites in Seismic Zones 3 or 4 also require steel reinforced cast-in-place walls.

The presence of interior walls also helps determine the need for reinforcement. Interior walls help brace foundation walls against lateral forces, so less reinforcement is needed. If the height of the backfill is more than 8 feet, the distance between interior bracing walls exceeds 12 feet, or the potential exists for high moisture content in the backfill, the reinforcement must be designed by a structural engineer.

Place the reinforcing steel rods, or rebar, in the wall cavity after the exterior forms are in place. Make sure that the steel reinforcement does not block the alignment of the wall ties with the interior set of forms. The distance between the reinforcing steel and the interior face of a 6-inch-thick foundation wall should be a minimum of 1 inch and not more than 2 inches. For thinner walls, the reinforcing steel usually is centered in the wall.

To install the rebar, align the interior wall forms with the exterior forms to allow the wall ties to be inserted easily. Cut plywood for the corners to compensate for the thickness of the wall. If the foundation walls are 8 inches thick, the length of the forms for the inside face should be reduced by 8 inches for inside corners. The length of the forms for the outside corners should be increased by 8 inches. The length of the form for the inside of a straight wall usually is shorter by two wall thicknesses than the length of the forms for the outside of the wall. Next, install the interior side walers and any necessary vertical or diagonal bracing. Be sure to secure the

FIGURE 7-5. Control joints divide walls and slabs into smaller sections to guide the formation of cracks caused when concrete expands and contracts.

Locating Control Joints

Typically, control joints in slabs on ground should be spaced 10 to 15 feet apart, making concrete sections as nearly square as possible.

If the sections cannot be square, the longer sides should never be more than 1 1/2 times the length of the shorter sides.

Assuming good curing, the required joint spacing can be influenced by many things, including—

- Total water in the mix. More water may mean more shrinkage, thus the need for more control joints.
- Weather conditions. High temperature and low humidity may mean more shrinkage and the need for more control joints.
- Aggregate. The type and size of aggregate may influence shrinkage and thus the spacing of control joints, as shown in the accompanying table.

Maximum Spacing of Control Joints, in Feet

Slab thickness	Slump, 4"-5" Maximum-size aggregate Less than ¾"	¾" and larger	Slump less than 4"
4"	8	10	12
5"	10	13	15
6"	12	15	18

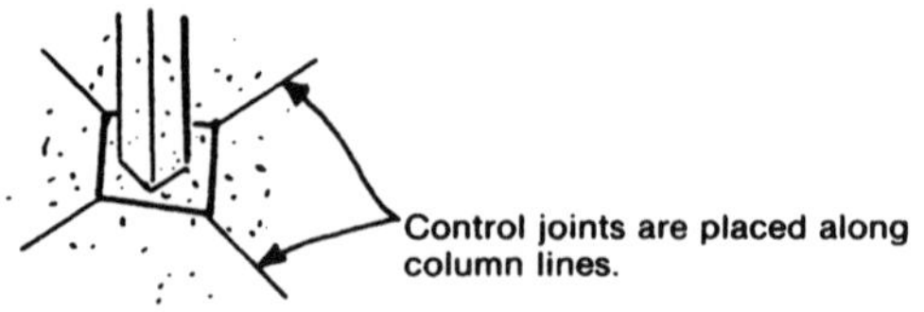

c. Control joints at columns.

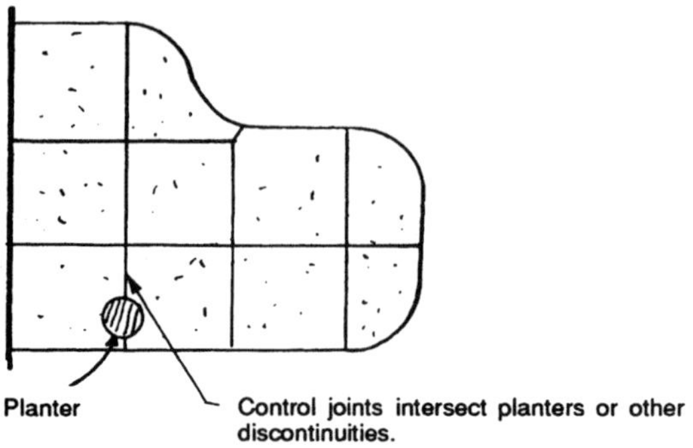

d. Control joints for a patio slab.

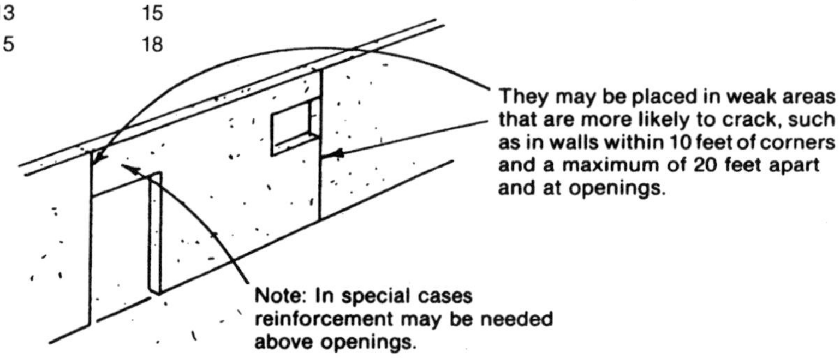

e. Control joints in a concrete wall.

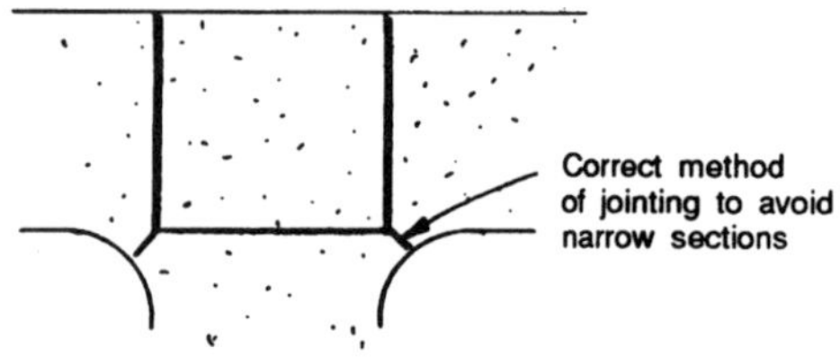

a. Proper layout of control joints in driveway.

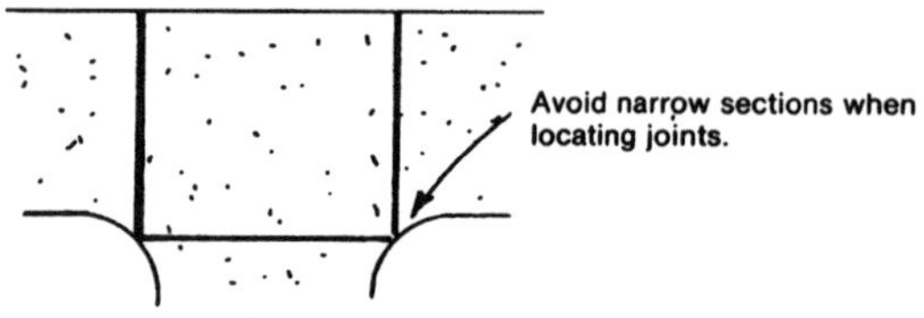

b. Improper layout.

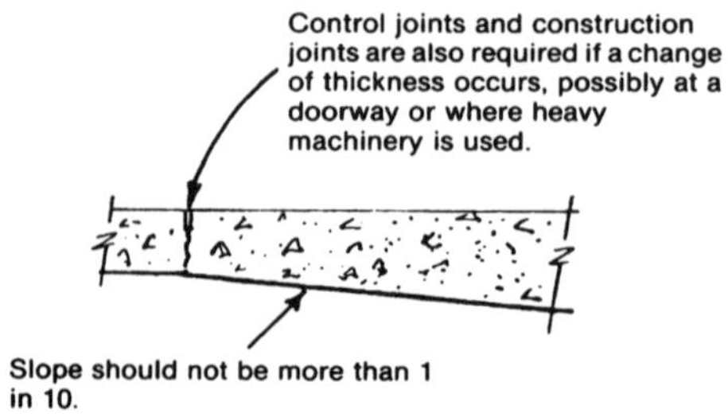

f. Control joint at a change in slab thickness.

Source: Residential Concrete, *3rd ed., Home Builder Press, National Association of Home Builders, Washington, D.C.; 1999.*

walers at the corners to prevent the forms from deflecting.

Creating Basement Foundations Using Manufactured Forms

Builders also use manufactured forms to create cast-in-place concrete walls. Forms are manufactured in many sizes. Most are 2 feet wide and available in heights of 3, 4, 5, 6, and 8 feet. Fillers are available in widths ranging from 2 to 22 inches in 2-inch increments. Several manufacturers produce forms designed specifically for residential basements, but virtually all manufactured wall forms can be modified for this use.

Manufactured systems generally include a smooth-faced material attached to a metal framework, usually made of steel or aluminum. The smooth-faced material may be plywood, a laminate material that has a very smooth, nonporous surface, or *plyform*, a type of plywood with a specially coated face designed for concrete forms. This framework provides both vertical and horizontal support, points for bracing, and a way to secure one panel to the next. Manufactured forms for corners and columns allow for fast, complete installation of a wall system.

If handled carefully and pretreated with releasing agents before the concrete is cast, most manufactured forms can be used at least 20 times before their facing surfaces must be replaced. The form frames can last indefinitely. Specific performance data for various types of manufactured forms are available from the manufacturers. Most manufactured form systems are available for either purchase or rent.

Installing the Forms

The manufactured forms are aligned with a wooden sill of 2×4s nailed to the footing to mark the perimeter of the walls. If the concrete in the footing is fresh, hand-nail the lumber to the footing using concrete or cut nails. A powder-driven nailer will work when the concrete has cured to a hardness that will not accept hand-driven nails. Powder-driven tools use gunpowder to fire nails into steel or hardened concrete and should only be operated by qualified workers.

Set the exterior wall forms first, attaching them to the sill to prevent the framework from *blowing out*. Join each form to the next using the integral attachments provided by the manufacturer. After the form for the exterior side is in place, place blockouts for doors and windows, insert the wall ties, and attach them to the interior wall forms. As with site-built forming systems, reinforce any headers above doors and windows with rebar.

Manufactured forms are available in a variety of heights and widths. The widths usually are measured in 2-inch increments. Because the system of measurements used in manufactured forms can limit their applicability to some designs, any necessary modifications to the design of a home's foundation should be made before the footing is cast. The most common forms are 2 feet wide, so the most efficient designs are those with wall dimensions based on 2-foot increments. Narrower forms may be required to allow the ties to align properly, but a 2-foot scale keeps sizing to a minimum and speeds up the setting of the forms.

Most manufactured forming systems require some horizontal bracing, which generally is placed after the forming sections are fastened together. Each manufacturer's system is designed for use with specific ties and bracing and has its own variety of clamps that attach to an exposed portion of the wall tie and secure the bracing in place. Any deviation from the manufacturer's requirements can create a dangerous situation, such as a blowout, and other potential problems with the cast and should not be attempted. An assortment of manufactured forms is shown in Figure 7-6.

As with site-built systems, fasten the reinforcing steel in place after the outside forms are in position. Keep the steel at least 1 inch away from the inside of the forms. The effectiveness of the rebar is compromised if the concrete does not surround it. The concrete cover also protects the rebar. Normally, the maximum size of the aggregate used in concrete walls is not larger than 3/4 of the smallest distance between the rebar and the form. For example, for a 1-inch clear distance, 3/4-inch aggregate is the largest aggregate that should be used in the concrete mixture.

Once the reinforcing steel is in place, set the forms for the interior of the wall, align them with the snap ties, and then fasten and brace them in place. Placement of the concrete can begin after all the forms are fastened together and all bracing is secured.

FIGURE 7-6. Manufactured forms are an alternative to site-built forms.

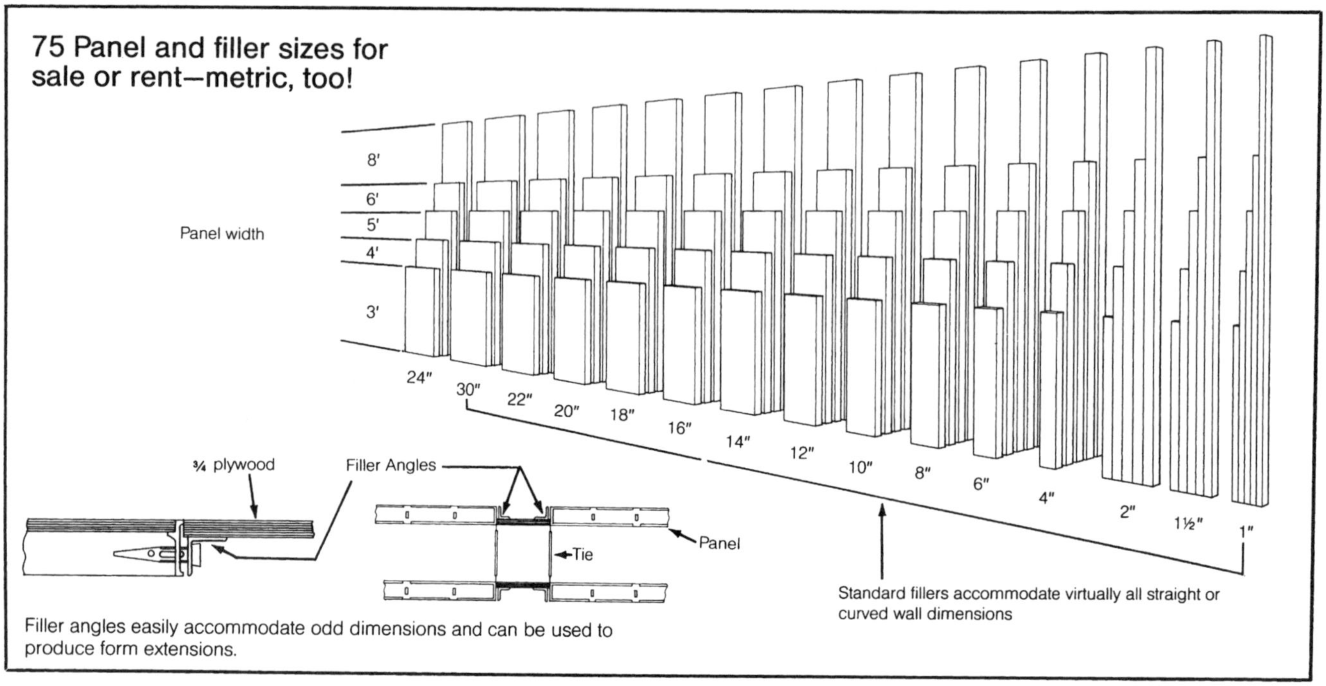

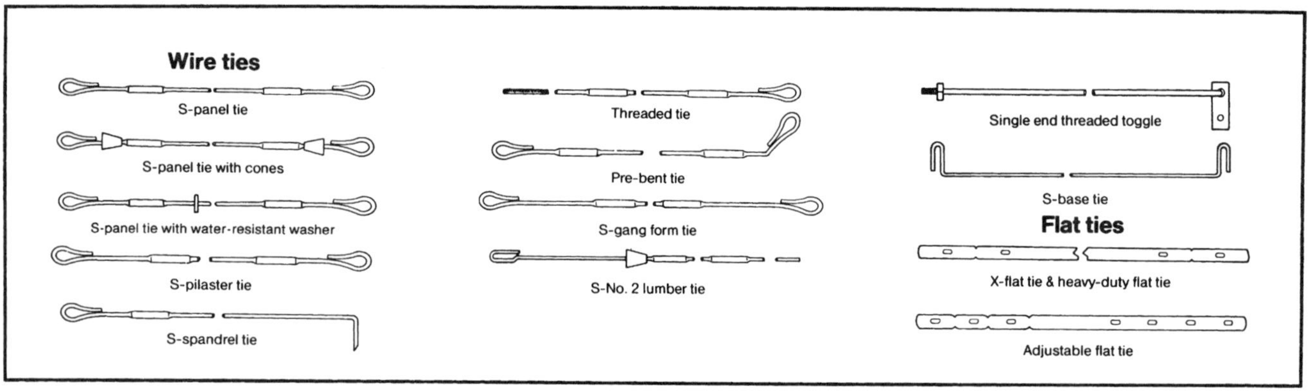

Source: Symons Corporation, Des Plaines, IL.

Placing the Concrete

Follow the manufacturer's recommendation regarding the *rate* at which the concrete should be placed. This rate usually is stated in number of vertical feet per hour. Most manufactured forms can be filled in a series of 1- to 2-foot lifts, with each lift being mechanically vibrated. See Chapter 5 for more information on lifts and vibrating techniques.

After the cast is completed but before the concrete begins to set, place sill plate anchors so the sill can be attached. Embed them in the concrete at intervals that meet code, usually 6 feet on center with one not more than 12 inches from each corner.

Do not remove the forms until the concrete has reached an adequate strength. The time required depends on the ambient air temperature. Retarding and accelerating admixtures present in the concrete mix also can affect curing time. In most cases, the forms can be removed after 24 hours. When the temperature falls below 40°, however, the curing time can double. Do not cast the foundation walls if the temperature is expected to go below freezing within 48 hours after the casting is completed unless precautions are made to protect the walls. Blankets can be wrapped around the walls to retain the heat from hydration and protect the walls from freezing.

Creating Decorative Castings

When concrete is cast the exterior surface tends to "echo" the surface of the form. If the visible portion of the wall will not be covered with stucco or some other surface, a decorative finish can be cast into the surface. Several manufacturers fabricate forms and form liners that attach to the inside surface of the form that simulate a patterned impression when removed, such as brick, tile, stone, or a geometric design. If liners are used they must be compatible with the forming system. As with all forms, apply a releasing agent before placing the concrete. If the surface of the wall will be visible, use special form ties that do not distort the decorative pattern. Decorative form liners and their resulting patterns are shown in Figure 7-7.

Curing Cast-in-Place Walls

After concrete is placed the water in the mix chemically reacts with the cement in a process referred to as *hydration*. This reaction forms the adhesive that binds all the aggregates together. The reaction continues after the concrete hardens, until the concrete gains its ultimate strength. For the hydration to be complete water must remain in the freshly placed concrete as long as possible. *Curing* is the process that holds the water in the freshly cast wall after the forms are removed.

FIGURE 7-7. Form liners cast a decorative pattern in the concrete, such as brick, tile, stone, or a geometric design.

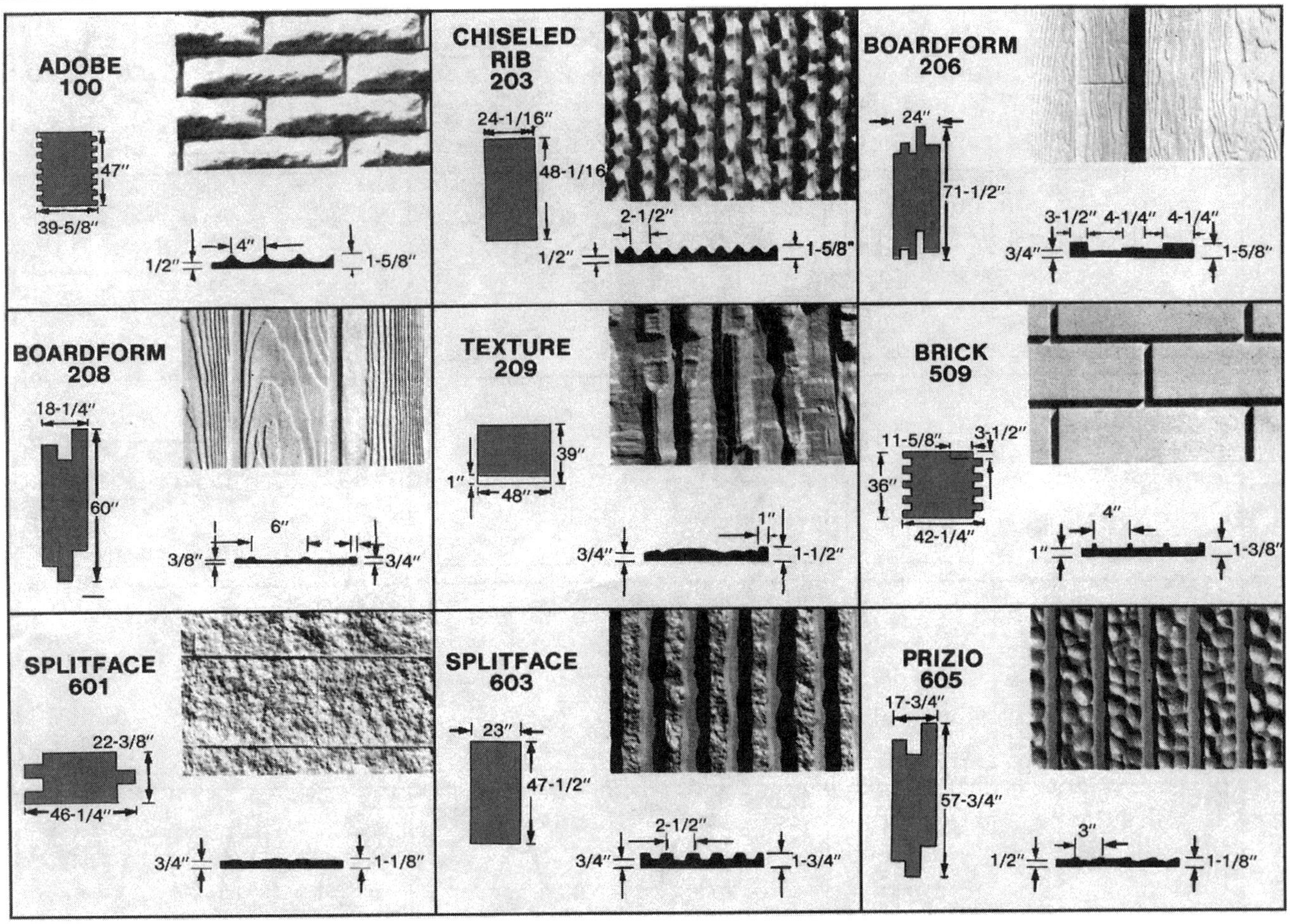

Source: Aluma Enterprises Inc., Toronto, Ontario, Canada.

Curing is most easily accomplished by spraying the wall with a curing compound after the forms are removed. This chemical forms a thin layer over the concrete and retards the evaporation of water, letting hydration continue and allowing the concrete to gain its designed strength. Do not neglect this important procedure.

Casting Stemwalls

Cast-in-place concrete walls used in slab-on-grade systems or for crawl spaces are referred to as *stemwalls*. Stemwalls rest on footings cast in trenched earth or forms and are, in most cases, about 20 inches tall. The top of the stemwall becomes the base for the wooden crawl-space structure. The stemwall should provide a minimum clearance of 18 inches between the bottom of the floor joist and the ground. The exterior finish grade can be higher than the ground within the crawl space, but the sill of the structure should be at least 8 inches above the finished grade of the lot. A stemwall for a slab-on-grade foundation must be built high enough for the finished floor of the structure to be at least 8 inches above the finished grade.

Because the in-place height of a concrete stemwall is only a few feet, the pressure exerted by the freshly placed concrete is significantly less than that created by a full-size basement wall. Because there is less pressure, less bracing is necessary, less material is required, and the overall process is less time consuming.

FIGURE 7-8. Secure the top of the stemwall forms using brackets.

Installing the Forms

After an appropriate footing is cast, outline the stemwall on the footing with a chalk line. Hand-nail *spreader cleats* to the footing using 1-inch concrete nails. Spreader cleats are metal clips that have slots that hold 2×8-foot sheets of 3/4-inch-thick plywood or plyform apart, creating a form for the stemwall. Use the same type of bracket to secure the tops of the forms. Figure 7-8 shows a stemwall being formed. Next, mark the top of the stemwall using an optical level and a tape. After the level is marked at the corners, snap a chalk line to mark the level of the top of the stemwall along the interior of the forms.

If the stemwall is for a slab-on-grade project, attach a horizontal 2×4 next to the form with the top of the board aligned with the chalk line. The 2×4 creates a notch in the stemwall that supports the perimeter of the slab and, particularly for 2×4-exterior-wall construction, provides a smooth floor.

To form straight stemwalls, the upper edge of the stemwall forms must be held in place. This is accomplished most easily by placing 2- to 3-foot-long stakes from the top of the form diagonally outward to the ground. Align the walls using the string line method for aligning site-built forms described previously in this chapter.

Venting the Crawl Space

Moisture under the crawl-space structure can damage the framing. To allow moisture from the crawl space to escape, most codes require vents. Place vent blockouts, usually 8×16-inch boxes, in the forms so that they align with the top of the walls. The boxes create voids in which a vent can be installed later. Most building codes require that a minimum of 1 square foot of ventilation be provided for every 150 square feet of crawl space area. If a 6-mil polyethylene sheet or a 2-inch topping slab covers the area beneath the slab and acts as a vapor barrier, the amount of ventilation can be reduced to a ratio of 1/1,500.

Because the screen mesh in the vents limits the flow of air, the *net free area* or unobstructed venting capacity of the vent should be marked on the vent. The net free area does not refer to the vent's actual size, but instead to the vent's capacity for airflow. Vents are available that can be closed during the winter to keep some warmth in the crawl space and help keep pipes from freezing and opened in the summer to vent humid air. Most vents have a net free area of about 1/2 of a square foot.

Some codes also require that a blockout for an access crawl hole of at least 18 inches by 2 feet be placed in the stemwall.

When creating a stemwall for a structure that has a bay or bow-bay window, stop the forms at the edge of the window. The shape of the window is easier to cast when the slab is poured.

Use multiple layers of 1/4-inch-thick forms to produce circular stemwalls. This is very time-consuming, taking three to four times as long as straight walls to form. The associated cost must be considered when the project is estimated.

Placing the Concrete

Once the forms are in place, the stemwall can be cast using an appropriately designed mix. In most cases 3,000 psi concrete placed at a 4-inch slump is required. Place frame-anchoring devices into the surface of the fresh concrete before it sets.

For a smooth floor under exterior doorways, remove the top 4 inches of concrete from the stemwall under the door openings before the concrete hardens completely or block out the area before the stemwall is cast. Prior to placing the concrete on the floor, place a board even with the top of the stemwall along the exterior side to create a form for the slab before casting the floor. When the floor is cast, the concrete will cover the stemwall and fill the void. This provides a smooth floor under the door's threshold.

Chapter Checklist

When using any of the described systems to erect a cast-in-place concrete wall, ask the following questions before and during the construction process.

- Is the perimeter marked on the footing and is it square?
- Will the foundation wall rest on the center of the footing?
- Is the surface of the form clean, smooth, and free of defects that might leave an undesirable imprint on the surface of the cast wall?
- Is there an adequate number of wall ties at the site?
- Is any reinforcing steel in the way of the proper placement of the wall ties?
- Are all blockouts for windows and doors properly placed and braced?
- Is the top of the cast-in-place wall marked on the inside of the forms?
- Are chamfer strips for the control joints in place, 10 feet from the corners and every 20 feet?
- Are the forms braced to prevent bowing and blowouts?
- Are the forms properly aligned?
- Has the interior of the forms been coated with a releasing agent?
- How fast can the forms be filled without putting additional pressure on the wall ties and forms? What is the acceptable rate of pour?
- What is the slump of the concrete and is it free of or does it contain appropriate admixtures? Has the specified amount of water been added to the mix?
- What is the air temperature? Are blankets available to protect the freshly cast walls from freezing?
- Is a vibrator available to vibrate the concrete during the cast?
- Are the correct numbers of specified sill plate anchors at the site?
- Is the crew adequate for the concrete casting, considering all contingencies?
- Is a proper work platform available? Standing on the top of the forms does not conform to Occupational Safety and Health Administration (OSHA) standards.
- Where can excess concrete be placed?
- Has the concrete in the walls reached adequate strength to remove the forms? This usually takes at least 24 hours at temperatures above 50°. The time is longer when the weather is colder.
- How will the concrete be cured? Will forms be left in place? Will a curing compound be used?

8

Building with Insulating Concrete Forms

Concrete is a very dense product that transfers heat easily, but because of its density changes temperature slowly. This characteristic of concrete is called *thermal mass*. Because of its thermal mass, concrete is not the best insulator. A study by Dow Chemical found that up to 22 percent of a home's heating energy could be lost through basement walls that are not insulated. (*Handbook of Fundamentals*, American Society of Heating, Refrigeration and Air Conditioning Engineers, Inc.; Atlanta, GA; 1989.) When required by local building codes or when the basement will be finished to provide added living space, basement walls should be insulated.

There are several methods available that will improve the insulating quality of concrete walls. For example, metal or wood furring strips attached to interior concrete walls with a powder-driven tool can be used to create a framework for installing polystyrene foam insulation with a drywall board finish.

Insulating materials such as polystyrene foam sheathing also can be added to the exterior of a basement wall before the backfill is placed. The foam sheathing can be glued to the clean, dry exterior concrete with a compatible adhesive that will not dissolve the sheathing. As an alternative, some builders place the sheathing against the exterior wall and let the weight of the backfill hold the below-grade sheathing in place. The portion of the wall that extends above the grade is coated with a cementitious material like stucco or covered with panels designed to protect the foam sheathing from damage and to make the wall more pleasing aesthetically.

Other insulating systems use preformed shapes of various configurations and sizes that stack together and then are filled with concrete. These systems, commonly referred to as insulating concrete forms (ICFs) are a cost-effective way to erect a residence that complies with the Model Energy Code's (MEC's) insulation requirements for basements and crawl spaces.

Most of the products described in this chapter can be used to build more than foundation walls. Many can be used for all exterior walls and to insulate joists for multifloor projects.

Using Insulating Concrete Forms

ICFs can be used to erect homes with walls that have R-values ranging from R-18 to R-30, providing the homeowner with significant savings on heating and cooling costs. These systems are classified into three groups. The first group is the *flat* wall system, which uses flat sheets of insulating sheathing held in place by wall ties to form a flat cast-concrete wall of uniform thickness.

The second group uses manufactured foam blocks to cast the concrete into a matrix of horizontal and vertical concrete cylinders spaced no more than 16 inches on center. The cylinders usually are reinforced with steel dowels to increase the flexural strength of the wall. This group of ICF systems is divided into two types. Systems in the *screen* type produce horizontal and vertical cores that are isolated from each other. *Waffle* forming systems, the second type, create a matrix of horizontal cores with a thinner web of concrete that ties them together.

The third group, known as the *post-and-beam* ICF system, uses manufactured blocks to create an array of vertical cylinders on greater than 16-inch centers that are tied to horizontal cylinders. The different types of ICF are illustrated in Figure 8-1.

Several producers of ICF systems have developed bracing and scaffold systems for their products. Because the Occupational Safety and Health Administration (OSHA) requires fall protection when workers are more than six feet above the ground, these systems may meet this OSHA requirement. The bracing usually is modular and can be assembled and dismantled quickly. The scaffolding segment of the system also expedites the placement of the concrete into the forms. The workers are on a stable platform and not wasting time moving ladders around the foundation. A bracing scaffold system is shown in Figure 8-2.

Local building officials and the model building code are beginning to recognize the use of ICFs as a sound construction method for residential foundations. To assist builders in obtaining local code approval of ICF systems the National Association of Home Builders Research Center, Inc. (NAHBRC) under sponsorship of the U.S. Department of Housing and Urban Development (HUD) published *Prescriptive Method for Insulating Concrete Forms in Residential Construction*, National Association of Home Builders Research Center, Inc., Upper Marlboro, MD; 1998. The information in this text provides procedures for using ICFs in one- and two-family home construction, eliminating the need for engineering in most projects.

Installing Flat Forming Systems

Polystyrene insulation board is used as a forming material to create this group of ICF walls. Specifically designed nylon wall ties are used with either 2-inch-thick extruded insulative sheathing panels or 2-pound density-expanded bead board. Because the concrete adheres to expanded polystyrene insulation bead board this system should be used only when the forms will be left in place after the walls are cast.

Concrete does not adhere to extruded polystyrene insulative sheathing, however, so this product can be reused as a lightweight forming material. These forms are relatively expensive and easily damaged. They have a much shorter life span than sturdier forms made of metal or wood.

While the basic process for installing each of the systems in this group is the same, some manufacturers have developed proprietary wall tie systems. Contact the manufacturer of the system being used to obtain specific information on how the boards tie together.

Each manufacturer has specific recommendations for the installation of their products. The following information is based on techniques common to these processes. Follow the manufacturer's recommendations to assure the best installation of a specific system.

Use a plywood template to drill holes for snap ties in each 4×8-foot insulative sheet. Then slot the edges of the sheets to receive additional snap ties. Nail a 2×4 form guide along the line marked on the footing to indicate the exterior side of the foundation wall.

As an alternative, fasten the sheets of insulative board together to form panels and then place them on the footing. Hold the panels together by wrapping twine around the exterior caps of the wall ties.

FIGURE 8-1. Insulating concrete forms provide an alternative to constructing walls with standard forms.

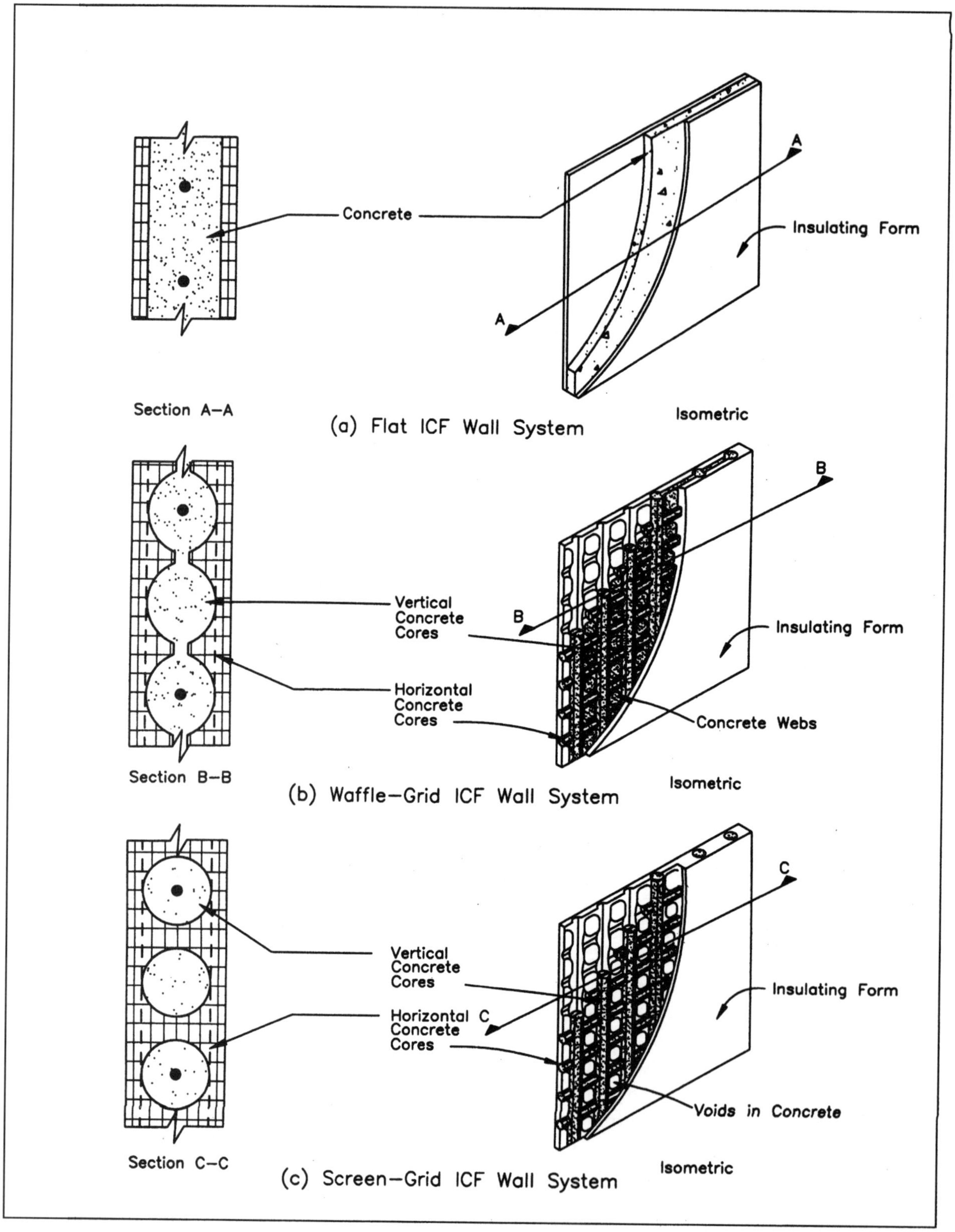

Source: Prescriptive Method for Insulating Concrete Forms in Residential Construction, *National Association of Home Builders Research Center, Inc., Upper Marlboro, MD; 1998.*

FIGURE 8-2. A bracing scaffold system enables workers to walk on a stable platform.

Source: Courtesy of Fold Form LC, Sioux City, IA.

FIGURE 8-3. Each manufactuter provides an in-wall bracing system, such as this Lite Forms bracing.

Source: Courtesy of Fold Form LC, Sioux City, IA.

If reinforcing steel is needed, place it before the forms are installed or attach it directly to the nylon wall ties. Vertical in-wall bracing of reinforcing steel, placed at 4-foot intervals, prevents the foam panels from deflecting while the wall is cast. Each manufacturer provides an in-wall bracing system. If the in-wall system is not used, contact the manufacturer for an exterior bracing design. Whichever system is used, the forms must be braced while the wall is cast. In-wall bracing for Lite Forms is shown in Figure 8-3.

Installing the Forms

Begin installing the forms at the center of each wall and move toward the corners. Press the metal plate, which is similar to the plate used to secure roof truss members together, into the sheets of foam to connect the individual sheets together. The sheets can be cut to adapt to any foundation dimension. Special metal corner straps available from the manufacturer attach to the wall ties and provide the necessary corner bracing.

Hold the tops of the forms in place using a ladder built of two parallel 2×4s spaced by 1×4s placed at intervals of 4 feet. Secure a lateral bracing of 2×4s spaced every 8 to 10 feet on one side of the wall. Securely nail the lateral bracing to the ladder and firmly anchor it into the ground. ICFs being set and braced are shown in Figure 8-4. The flexibility of ICFs also can be used to design special features, such as the curved wall shown in Figure 8-5.

Fasten blockouts for doors and windows to the foam before placing the concrete. To make the blockout rip a 2× to a width equal to the void between the interior and exterior form and slide it into place between the forms. Securely fasten it with nails or screws. Some manufacturers suggest that screws be driven through the 2× so that the end of the screw enters the void where the concrete will be placed. When the concrete is placed, the screws will securely anchor the blockout to the concrete. As with other forming systems, place horizontal steel reinforcements in any headers created above windows or doors. To facilitate the removal of the blockout after the wall is cast, coat the blockout materials with a releasing agent that keeps the concrete from bonding with them. A blockout for a window or door is shown in Figure 8-6.

Blockouts for brick ledges or beam pockets also should be placed prior to casting. Fasten pieces of dimensional lumber to the inside of the form with screws or nails driven through the forms. If a taller blockout is required, attach sheathing to lumber pieces to form a three-sided box. Then place the box in the form with the closed side toward the center of the void.

FIGURE 8-4. Set and brace insulating concrete forms before placing the concrete to prevent blowouts.

Source: Foam Form Systems, L.L.C., McHenry, IL.

FIGURE 8-5. The flexibility of insulating concrete forms makes features, such as this curved wall, easy to incorporate into home designs.

Source: Courtesy of Fold Form LC, Sioux City, IA.

Placing the Concrete

When working with polystyrene foam forming systems, place 3,000 to 3,500 pounds-per-square-inch (psi) concrete with a maximum of 3/8-inch aggregate at a maximum 4- to 6-inch slump in 3- to 4-foot lifts at a rate of approximately 8 vertical feet per hour. Vibrate the concrete using a 3/4-inch concrete vibrator. This small vibrator allows the mechanism to probe into the forms while minimizing the chances that the reinforcing steel will be misaligned. If the concrete is not vibrated properly, the concrete may not completely fill the forms, creating voids around the reinforcing steel and possibly causing a blowout.

Because the potential for blowouts exists with foam forming systems, blowout clamps are recommended. A clamp can be built before the wall is cast using two 24-inch 2×4s drilled and held together by a threaded, 3/16-inch-

FIGURE 8-6. Fasten blockouts for doors and windows to the foam before placing the concrete.

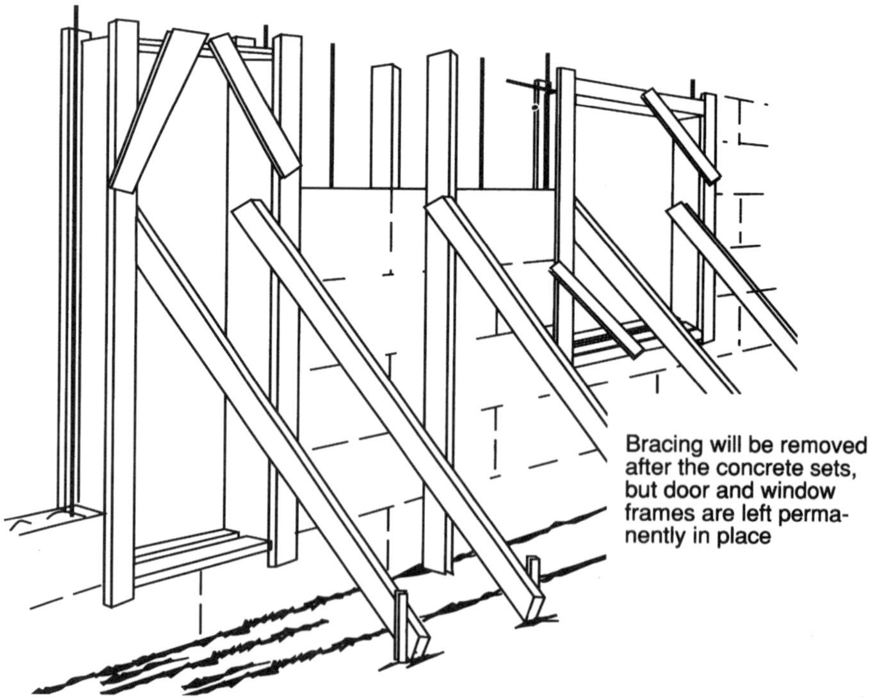

Source: Drake E. Gastineau.

diameter rod. Insert the clamp through the form at the point of the blowout to hold the damaged form together. As with all cast-concrete walls, place framing anchor bolts after the casting is completed.

One advantage of ICFs is that they protect the concrete as it cures. Even in freezing weather, the heat generated by the concrete during hydration is captured by the insulation, protecting the fresh concrete from damage. Even the exposed concrete along the tops of the forms may be protected. Because of the insulating qualities of these forms it is possible to cast foundation walls at temperatures that are below those recommended for traditional forming materials. The insulative sheeting also decreases the amount of water that evaporates out of the foundation wall and aids the curing process.

With insulative sheeting, concrete usually cures to sufficient strength 48 hours after placement. In cold weather the forms should be left on longer. Concrete must cure for at least three days at or above 50° F. Removing the forms before this time is not recommended or beneficial.

After the concrete cures, remove the forms or leave them in place to provide insulation. When removing the forms, first remove the large plates on the snap ties so that the sheathing can be pulled away without being damaged. Then snap off the remaining portion of the wall ties at the concrete's surface. The forms usually can be cleaned and reused twice. If the forms will be left in place, apply a latex-based moisture coating to all below-grade surfaces and a durable protective coating to all above-grade surfaces. Many solvent-based coatings will dissolve the foam, so it is important to check that the coating is compatible with the product.

Cutting in Mechanical Systems

Placing plumbing or electrical lines on the interior of the foundation wall is simple. Cut the sheathing with a router or a specialized knife that uses a hot wire as the blade. The hot wire melts the foam and the groove created provides channels for electrical conduit and plumbing pipes. Apply drywall over the sheathing by attaching the drywall to the wall tie plates with drywall screws.

Casting Basement Walls with a Waffle or Screen ICF System

Several companies manufacture stay-in-place ICFs made with expanded polystyrene foam. Typically, the foam wall forms are 9 to 12 inches thick, 12 to 16 inches tall, and 48 inches long. Some forms come in metric sizes. Two shapes typically are available, one for turning corners and the other for the main body of the wall. The concrete is placed into a hollow cell within the forms.

Some ICF manufacturers also make an integrated footing unit for their system. The top of the footing unit usually is at least twice as wide as the wall units and the bottom generally is bell shaped. This configuration transfers the load of the foundation wall over a wider footprint and eliminates the need for a separate footing. These units can reduce production time because forms for the footing do not need to be built or stripped. A footing unit is shown in Figure 8-7.

Installing the Foam Forms

While the components of each manufacturer's system differ, the installation procedures are similar. The wall systems usually are installed on 18-inch-wide footings. Most manufacturers suggest that a steel dowel be placed into the footing so that it projects up and into the first course of polystyrene blocks to tie the footing and wall together. The dowel placement should conform to the core of the particular system used.

Snap a chalk line on the footing to outline the perimeter of the structure. Nail two 2×6s together to form L-shaped braces and position a brace on each corner. Plumb the corner braces and secure them in place with 2×4 bracing that runs diagonally from the top of the corner brace to the runner attached to the footing. Then secure a 2×4 so that it runs along the exterior of this line. Or apply a bead of foam to one side of the line. The bead of foam acts as glue to bond the foam block to the footing.

Trim the bottom of the first coarse of blocks to compensate for irregularities in the footing. Cut the polystyrene blocks to the desired length to enable the blocks to fit together. Begin placing the blocks at the corners and move towards the center of the walls. After the first course is placed and the walls are squared, install a second 2×4 to the footing along

FIGURE 8-7. An integrated footing unit with a bell-shaped bottom may eliminate the need for a separate footing.

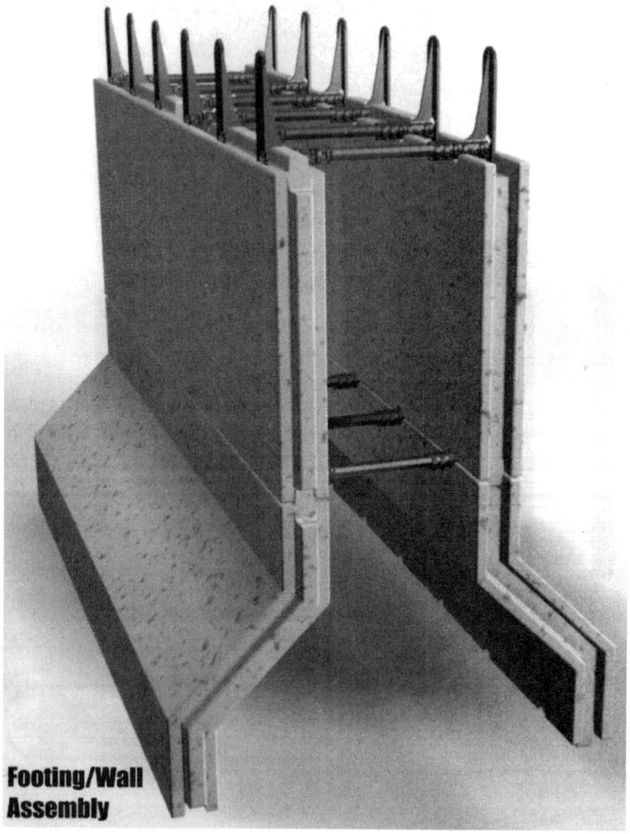

Source: GlobalTECH Building Systems, Inc., Columbus, OH.

the interior of the wall to hold the base of the blocks securely in place.

Then place the remaining courses along with the necessary horizontal and vertical reinforcing steel. For walls with the greatest strength, place both the vertical and horizontal rebar in the center of the cores created by the forms. Rest the rebar on chairs designed for the specific product being used. In some cases the vertical reinforcement can be installed after the final course of blocks are set in place.

Building Blockouts and Bracing

As with all systems, provisions must be made for windows and doors. Place blockouts made with dimensional lumber in the walls as the polystyrene blocks are placed. Brace the blockouts internally to prevent the wall from bowing from the weight of the concrete. Temporary 1×4s can be attached to the blockouts to form a pocket for the blocks. Brace the sides of the blockouts against the ground to maintain both the horizontal and vertical wall alignment.

Before the walls are cast brace the forms according to the manufacturer's requirements. This usually requires placing vertical 2×4 or 2×6 bracing along both sides of the wall every 8 to 15 feet. A diagonal member then is attached as a vertical brace. Attach one end toward the top of the wall, fasten the other end to a stake that has been driven into the ground. Align the forms using a string line or laser level. Site conditions such as high winds may require additional bracing.

Casting the Walls

Cast the walls using a minimum 2,500 psi concrete with a 4- to 5-inch slump. Any placement method will work, but because of the small spaces involved a pump with a small hose is recommended. A pump also minimizes the drop height, eliminating concerns about the segregation of the aggregates in the mix.

Be sure the concrete is placed at the proper rate for the forming system. Placing concrete too quickly can cause inadequately braced or assembled forms to "float" out of place. If this occurs, follow the manufacturer's procedures for correcting the problem.

To allow the entire depth of the form to be consolidated and reduce the possibility that the alignment of the reinforcing steel will be disturbed, vibrate the concrete using a 3/4-inch concrete vibrator to eliminate any voids or air pockets. As an alternative consolidation method, some manufacturers recommend rapping lightly several times with a mallet on a short length of a 2×4 placed against the form's ties. After the concrete is placed install anchor bolts or similar devices.

Because the forms do have a substantial insulating value, the heat from hydration should be enough to protect the freshly placed concrete from freezing. Still, covering and securing an insulating blanket or tarp to the top of the forms is a good preventive measure.

Finishing the Walls

Protect the exterior surface of the ICF materials after the wall is cured. Cover the below-grade portion of the wall with a latex-based moisture sealant and all exposed form materials with a wearing surface. Above the grade, the wall may be covered with stucco or brick supported on an integral brick ledge. A covering being applied to above-grade ICF material is shown in Figure 8-8.

Several building codes require that the interior surface of ICFs be covered with similar products or 1/2-inch drywall. To hang the drywall, some ICF manufacturers recommend installing furring strips to the concrete core of the wall with powder-driven nails. Other systems provide metal or plastic anchors as an integral part of the form. A router or a hot knife can be used to cut a channel in the foam for the installation of electrical conduit or plumbing pipes, as shown in Figure 8-9.

Placing Insulation Inside the Wall

Some builders incorporate polystyrene insulation board directly into the cast wall. One patented wall-tie system holds the insulation board in the center of the void between the sides of the form. The concrete then is placed into the forms, sandwiching the insulation between two parallel slabs of concrete.

The nonmetallic ties used with this system remain in place after the forms are removed, providing reinforcement to bind the concrete panels together with the insulation sandwiched between. Made of specialized fiberglass and vinyl ester polymer, these wall ties also minimize the transfer of heat and, unlike metal ties, are not affected by the chemicals found in many soils.

FIGURE 8-8. Cover exposed form materials with a wearing surface product.

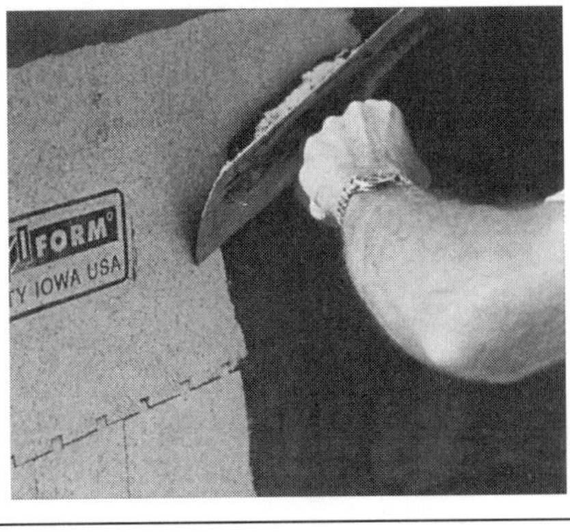

Source: Courtesy of Fold Form LC, Sioux City, IA.

FIGURE 8-9. Use a router to cut channels for mechanical systems.

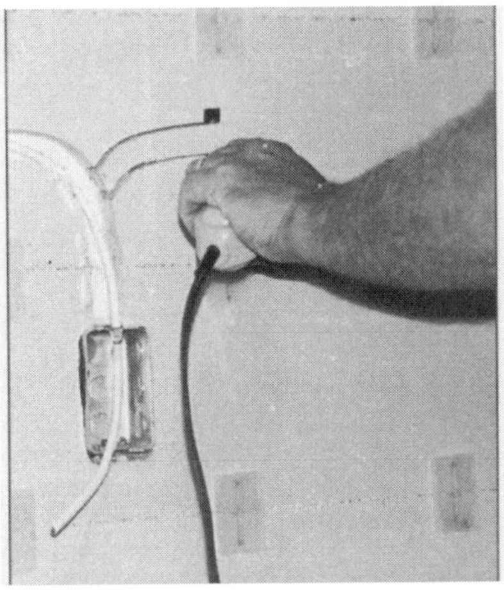

Source: Courtesy of Fold Form LC, Sioux City, IA.

This type of insulated wall protects the insulation from damage, overcomes the code requirement that foam sheathing be covered by a noncombustible coating such as drywall, and provides a thermal break in the wall. A vertical-pour system is illustrated in Figure 8-10.

Chapter Checklist

Plan the order carefully when using manufactured panels. Once an initial order is placed, incorporating changes becomes difficult and expensive, and can throw off the best-calculated schedule. Casting panels in site-built forms allows more flexibility, but changes are still cumbersome once the concrete is cast.

Consider the following questions when working with ICFs.

- Are adequate supplies on hand, including foam, sheathing or blocks, and accessories such as ties?
- Are the blockouts for doors and windows correct?
- If the blockouts will remain in place as nailers are they made from treated lumber?
- Are the walls straight, plumb, and properly braced?
- Is a repair kit available in case a form breaks?
- Has the correct psi concrete been ordered? Is the slump acceptable?
- What is the correct fill rate needed to prevent the lightweight forms from floating?

FIGURE 8-10. A vertical pour system is required for walls that incorporate an insulation sheet between two parallel slabs of concrete.

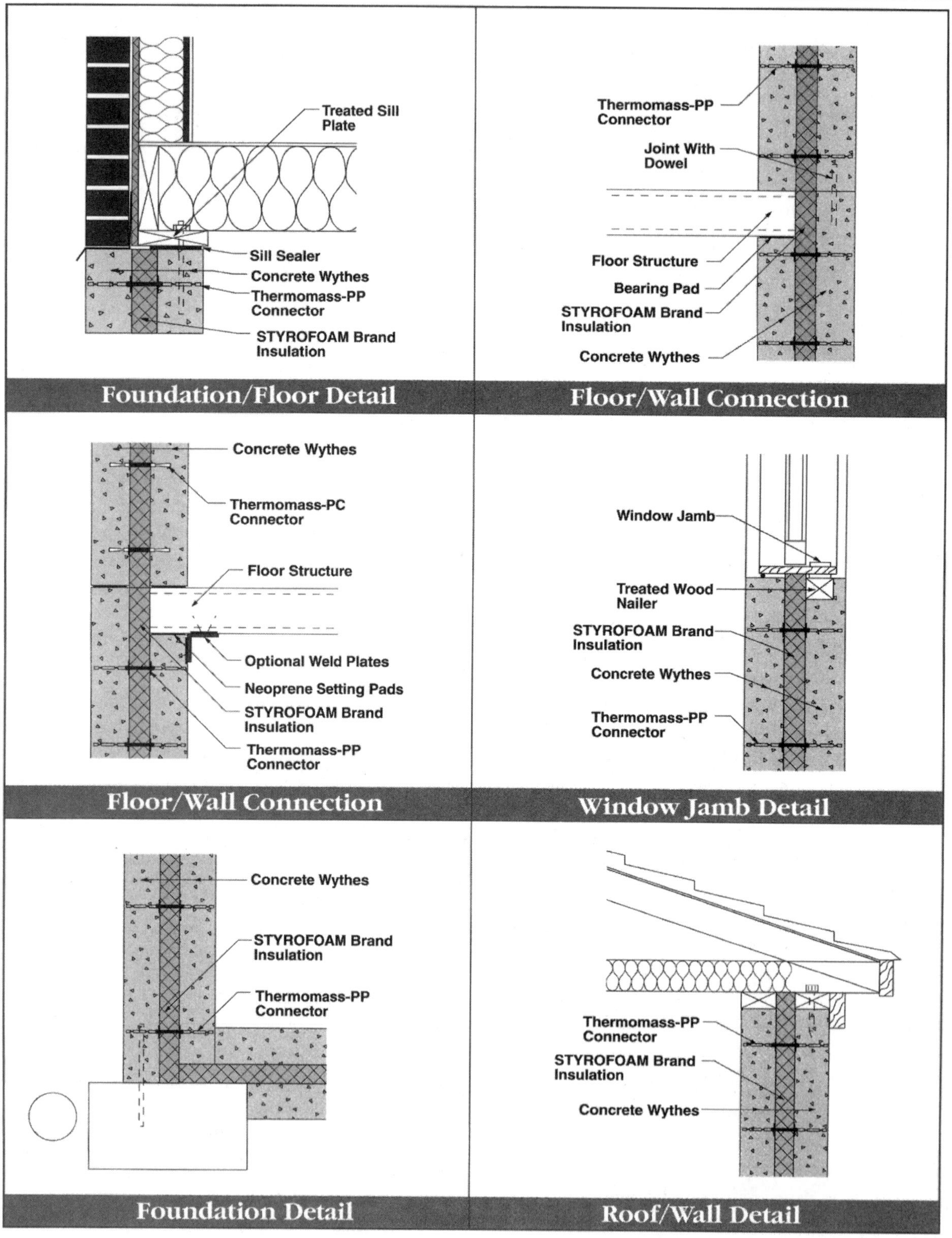

Source: Composite Technologies Corp., Ames, IA.

9

Using Precast Foundation Panels

Concrete wall panels also may be precast off site and, after curing, delivered and assembled into a complete foundation system. Most manufacturers make the panels for each job from an original or faxed set of drawings. Others stock wall panels in common lengths such as 10, 12, and 16 feet. Precast wall panels may or may not include steel reinforcement, depending on the requirements of the foundation design.

Precast panels offer several advantages. Because they usually are built offsite and delivered, the site can be prepared while the panels are being cast reducing construction time. Also, onsite weather conditions do not interfere with the production of the walls. Because the panels are produced at a facility that protects them from inclement weather, precast panels can be manufactured even when site conditions would prohibit the casting of concrete. The controlled conditions also help produce panels of consistent quality.

Several systems are used to create precast basement panels. Some use traditional slab casting principles while others use proprietary panel systems. Those using the traditional method start with a level base and form the edges of the slab. The slabs usually are cast 6 to 8 inches thick. Metal plates used to weld the panels together, as well as any required structural reinforcing steel and lifting anchors, are inserted in the form before the concrete is placed. The concrete is finished as if it was a slab on grade. Decorative patterns and exposed aggregate surfaces also can be applied to the surface. Figure 9-1 shows a forming panel.

Once manufactured panels are delivered they are hoisted into place with a lifting device or crane. Then the metal plates are welded together to connect the panels. Temporary bracing also is welded to plates cast into the panels. The seam between the panels is sealed with a high-grade caulking compound. For more information on using this precast system

FIGURE 9-1. Precast panel walls incorporate the required structural reinforcing steel, as well as lifting anchors and plates that are welded to hold the panels together.

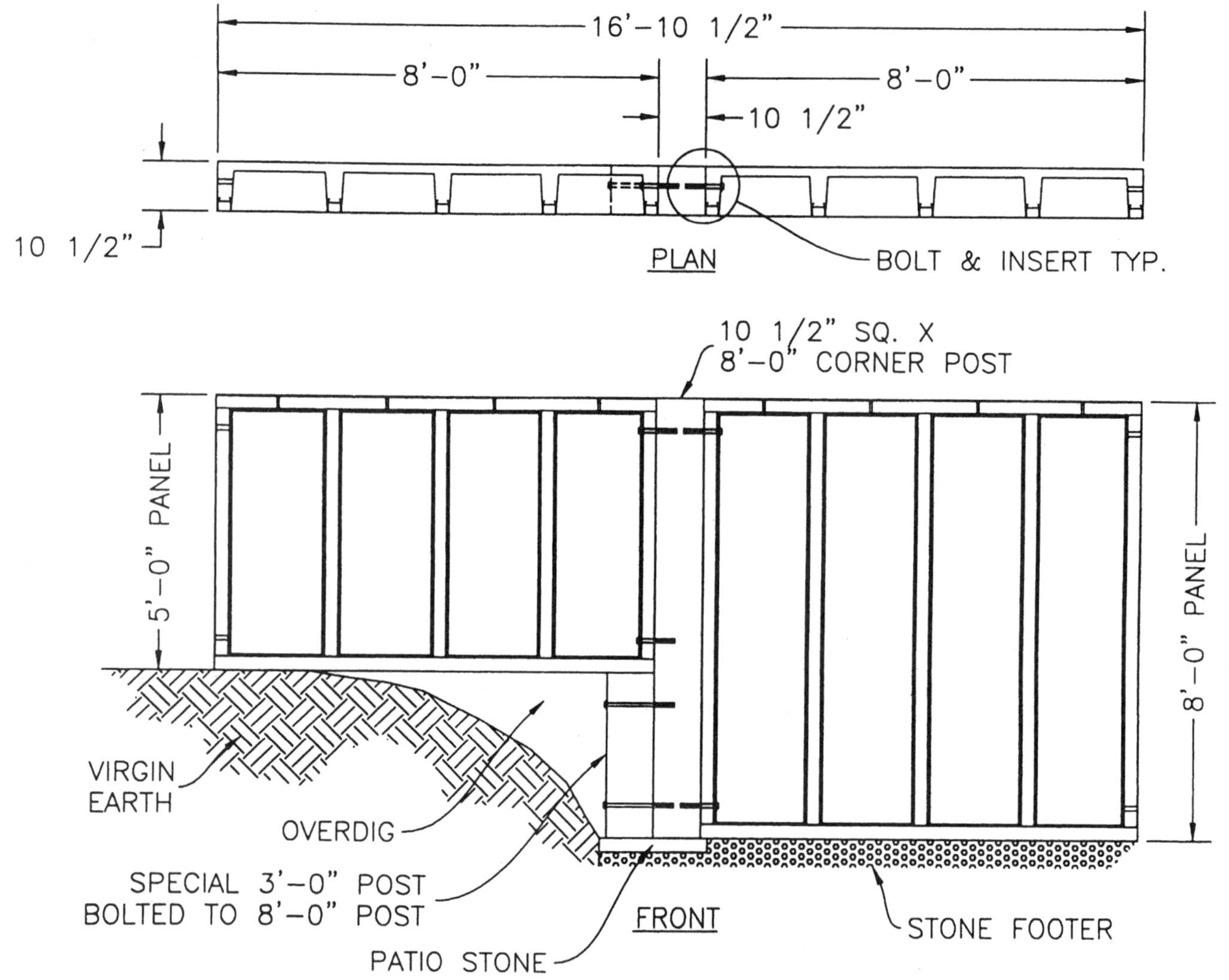

Source: Thermal-Krete USA, Kistner Concrete, East Pembroke, NY.

for residential work, contact the Multifamily Construction Advisory Committee of Illinois at 708-297-6704.

Using Structural Concrete Wall Panels

Thermal-Krete USA licenses one proprietary method used to create factory-manufactured foundation walls of high-strength concrete. The 10 1/2-inch-thick structural panels are available in 8- and 10-foot-tall panels. The panels are cast in molds that create a matrix of 2 1/2-inch-thick studs reinforced with 5/8-inch diameter #5 rebar placed 24 inches on center. The rebar is backed with a solid 2-inch-thick slab.

Another slab joins the ends of the studs to form a 3-inch-thick footer plate reinforced with two 1/2-inch diameter #4 rebars and a 2 3/4-inch-thick sill plate also reinforced with two #4 bars.

The interior edge of the stud is covered with a steel channel filled with expanded polystyrene (EPS) insulation. If a finished basement is desired, the drywall can be fastened to the insulation. A pair of erection anchors is cast into the top plate of the panel, as well as a series of holes that are cast in it for anchor bolts. Erection anchors also are cast in the footer plate. The studs at the end of each panel have a hole cast near the top and bottom through which a threaded bolt is passed to fasten the panels together. A drawing of a Thermal-Krete panel is shown in Figure 9-2.

Superior Walls of America also manufactures code-approved structural concrete panels that include concrete studs and a flat structural panel. The first step in this manufacturing process is the casting of concrete studs that are approximately 2 inches thick and 7 inches wide. The studs can be ordered in various lengths but for residential basements usually are made 93 inches long. An illustration of a Superior Walls panel is shown in Figure 9-3.

Into one edge of each stud is cast a 1×2 of chromated copper arsenate (CCA)-treated lumber. These wooden nailers allow interior finishes such as drywall to be attached easily. The other edge of the stud incorporates metal anchors that are used to connect the studs to the concrete face. The studs have uniformly spaced holes for the installation of plumbing and electrical systems.

FIGURE 9-2. Manufacturers license proprietary methods for factory-cast wall panels, as shown in this illustration of a Thermal-Krete wall panel.

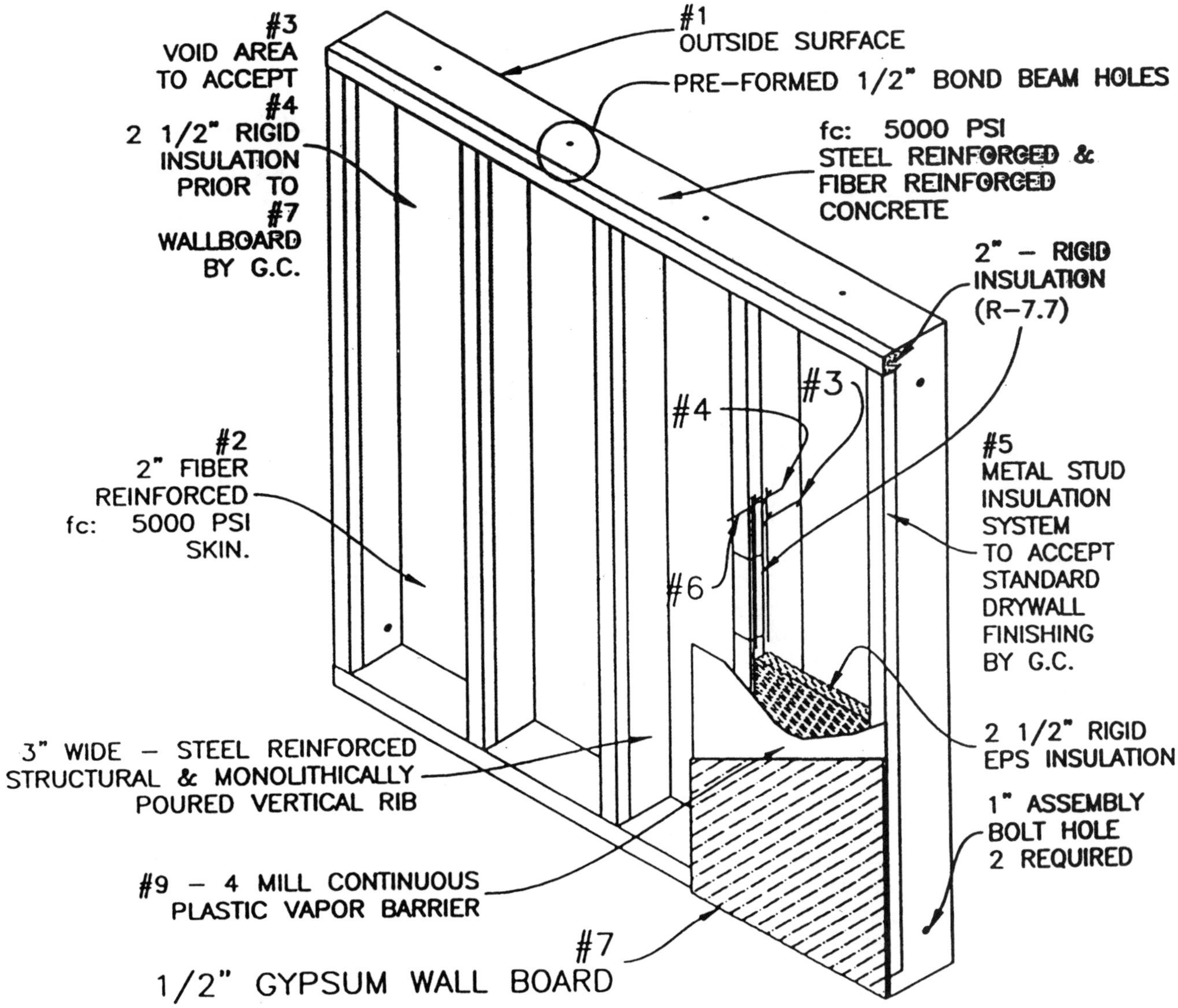

Source: Thermal-Krete USA, Kistner Concrete, East Pembroke, NY.

FIGURE 9-3. Wall panels manufactured by Superior Walls incorporate wooden nailers.

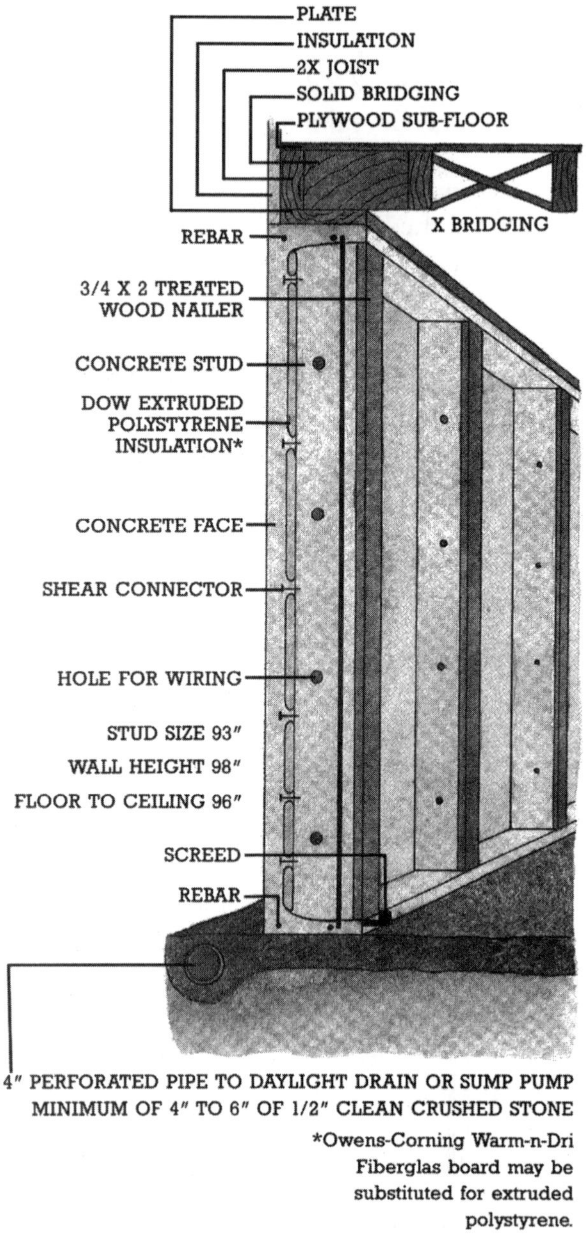

Source: Superior Walls, New Holland, PA.

The completed studs are spaced in a mold 24 inches on center, leaving a gap of 2 1/2 inches between the top and bottom of the studs and the edges of the mold. The wooden nailers face the bottom surface of the mold. Filler is placed between the studs. This leaves a gap between the top and bottom of the studs and the mold, forming a continuous beam that connects the tops and bottoms of the studs. These beams could be likened to the top and bottom plates of a conventionally framed wall.

Two runs of reinforcing rebar are placed inside the beam mold to enhance the flexural strength of the finished panel. Blockouts made of CCA treated lumber are placed in the form to create spaces for doors and windows. A sheet of polystyrene foam sheathing then is placed over the studs. Holes punched in the sheathing allow the stud anchors to pass through the foam. Any anchors and molds for fastening holes are placed next. Fillers that create required beam pockets also are placed, as shown in Figure 9-4.

A mixture of 5,000 psi concrete with a fiber-reinforcing additive, such as fibermesh, is placed in the form and in the beam pocket created between the tops of the studs and the forms. The resulting surface is finished using traditional methods.

Assembling Precast Panels

Although different methods are used to produce the various panels, the jobsite procedures for assembling the panels made by each manufacturer are similar. The depth of the basement is determined and excavated. Usually a slight slope is shaped into the bottom of the excavation to promote the flow of any surface water to the perimeter of the residence. A well-drained gravel footing is used to support the panels. The surface of the gravel footing must be level and uniformly compacted to bear the imposed load. Inadequate compaction of the gravel will allow the panels to settle unevenly, which destroys the integrity of the system. A perforated drainage pipe at least 4 inches in diameter should be placed on the interior side of the footing to promote drainage.

The panels are delivered and stored at the site or, if time and distance permits, trucked to the site and taken off the delivery vehicles and immediately placed in the excavated basement. If the delivery trucks must leave a paved surface, provide a temporary driveway.

Each panel is lifted into place with the use of a crane. The panels are attached mechanically using ties that pass through holes precast at regular intervals. Corners are formed with either precast L-shaped units or mitered corners. A wall being hoisted into place is shown in Figure 9-5.

Most installers start the Thermal-Krete system at one corner and finish one wall at a time. Installation

FIGURE 9-4. Fillers are placed in the precast walls to create beam pockets per builder specifications.

Inside Corner Beam Pocket

Left

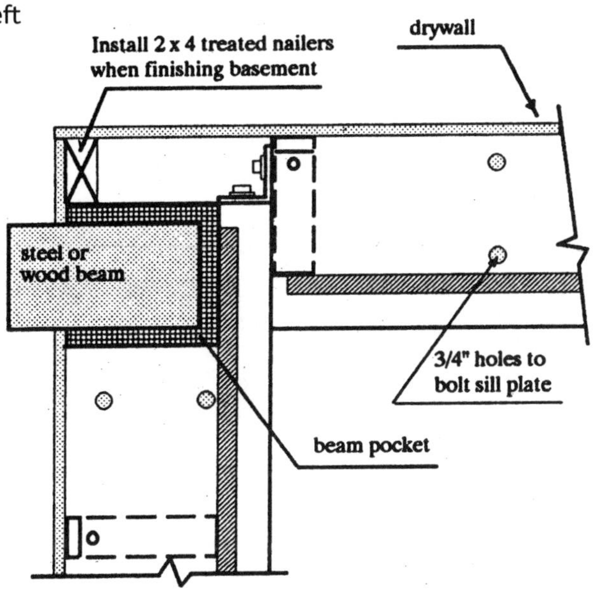

Right

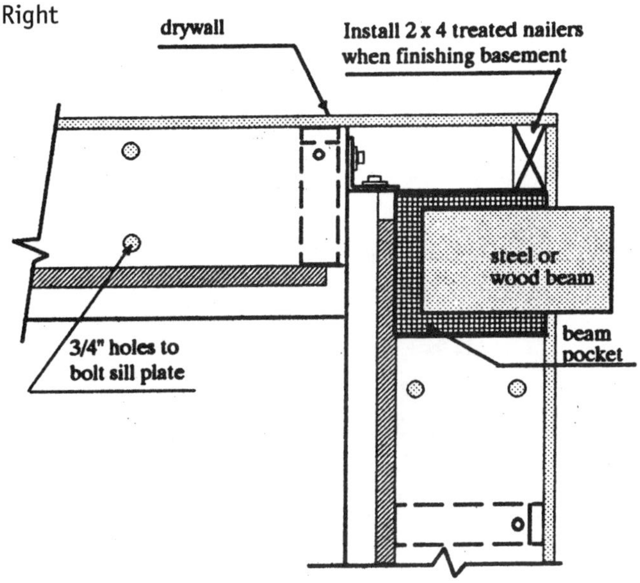

Standard Beam Pocket

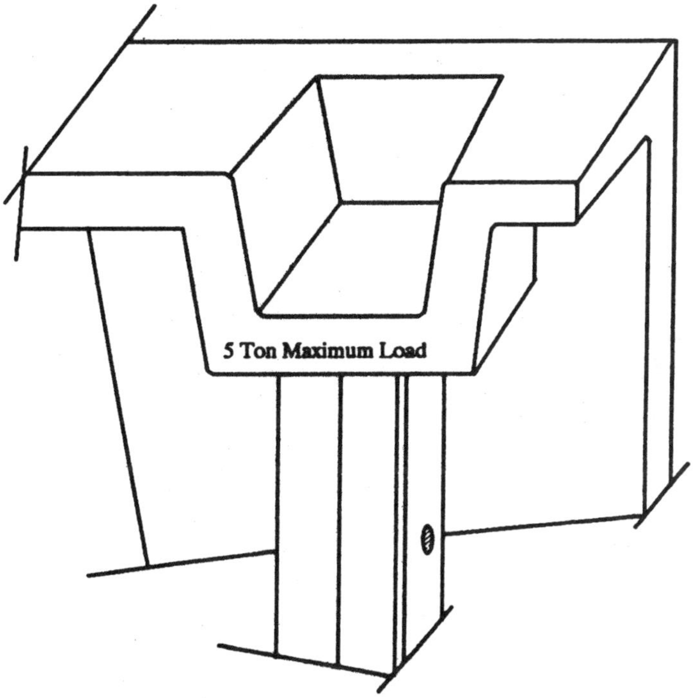

Source: Superior Walls, New Holland, PA.

of a Thermal-Krete corner and the associated bracing are shown in Figure 9-6. The corner units of the Superior Walls system must be joined together and the panels braced during the erection process. Apply two 1/2-inch beads of caulking to the abutting surfaces of each panel to form a watertight seal. Properly align the joints because the caulk, once set, permanently bonds the panels together. After the units are joined the weight of the walls is significant so it is critical that the panels be placed correctly. A corner installation of a Superior Walls system is shown in Figure 9-7.

As each panel is placed measure the distance from the corners to the end of each wall and the diagonal distance between these points. The diagonal distance should equal the square root of the sum of the other two sides. See Chapter 3 for additional information on using diagonal distances.

Attach a pressure-treated nailing plate to the top of the foundation walls after the walls are erected to allow for the attachment of a traditionally framed structure and place blocking between the floor joists at 10-foot intervals. The blocking keeps the top of the walls that run parallel to the floor joists from moving inward.

FIGURE 9-5. The precast walls are hoisted into place with a crane.

Source: Superior Walls, New Holland, PA.

Finishing the Walls

The nailers in precast wall panels make it easy to install interior finish surfaces such as drywall. Molded 1-inch holes in the panel studs allow the installation of all required electrical and plumbing lines. Insulation can be placed in the wall cavity.

Structural foundation panels are engineered to support an exterior facade of brick or stone. The facade can rest directly on top of the panels or on a brick ledge cast into the side of the panels. Incorporating a brick ledge into the design can be useful if the house will be set on a sloping lot and the exterior concrete surface of the foundation is not exposed. A brick ledge is shown in Figure 9-8.

FIGURE 9-6. Most installers of Thermal-Krete panels begin by placing and bracing one corner.

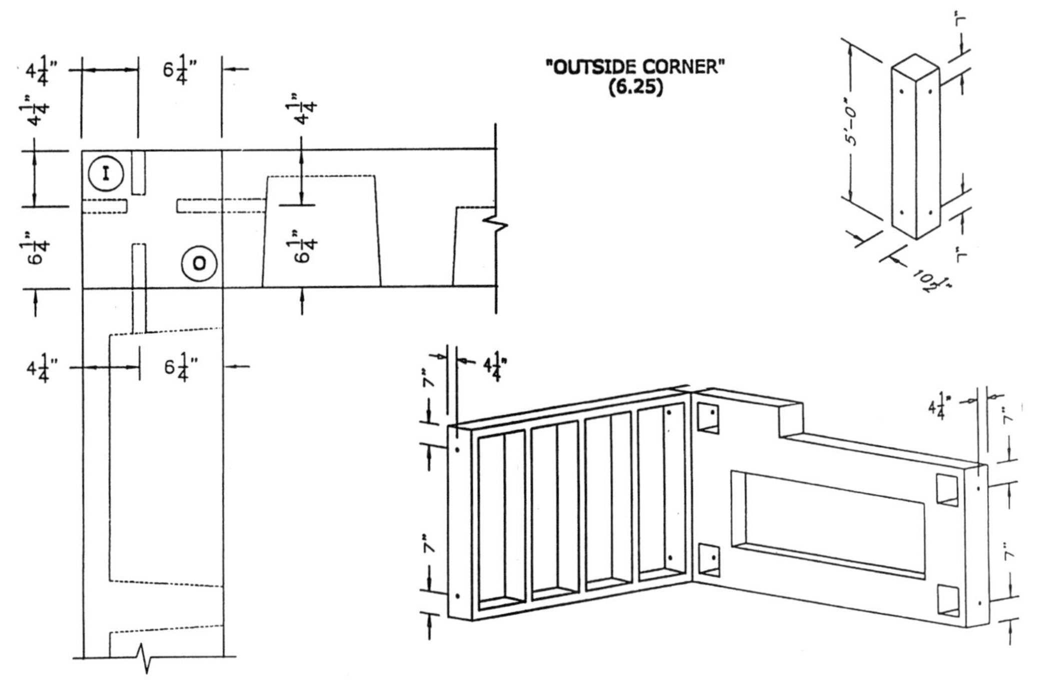

Source: Thermal-Krete USA, Kistner Concrete, East Pembroke, NY.

FIGURE 9-7. After aligning two walls, installers of Superior Walls panels caulk the corner joint permanently bonding the walls together.

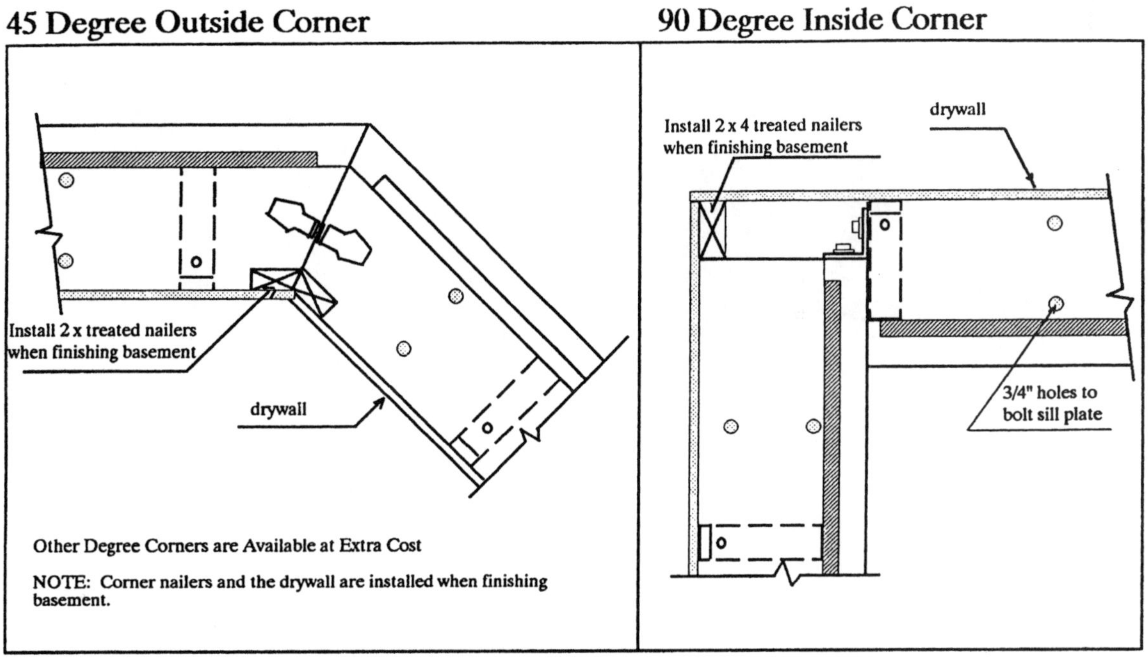

Source: Superior Walls, New Holland, PA.

FIGURE 9-8. Precast wall panels can be made with a brick ledge to support a masonry veneer.

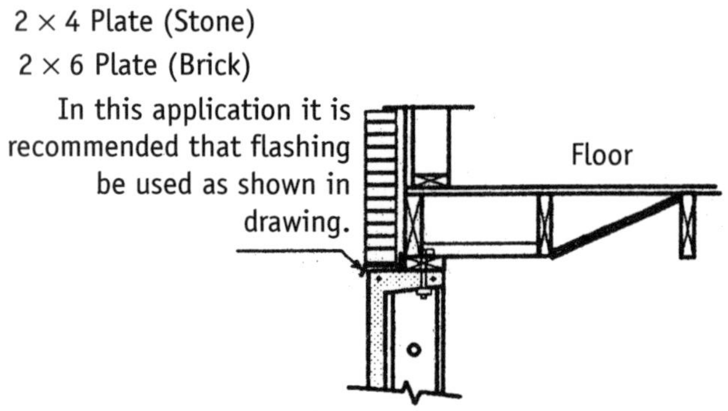

Source: Superior Walls, New Holland, PA.

Modifying First-Floor Framing

First-floor framing must be modified slightly on basements made of precast panels. Because the units are joined at several points along the wall, the panels might deflect slightly at the joints. Solid blocking placed between joists that run parallel with the wall will prevent this. Before the excavation is back filled, place the blocking no more than 8 feet apart and at the corners. The blocking must extend inward a minimum of at least 4 feet. This additional blocking is not required between joists that are perpendicular to the basement walls. A complete basement built with precast panels is shown in Figure 9-9.

FIGURE 9-9. Precast panels, shown in this illustration of a completed basement, can be customized to meet all design requirements.

Source: Thermal-Krete USA, Kistner Concrete, East Pembroke, NY.

Chapter Checklist

When using precast panels, consider the following questions.

- Is a surface available that can accommodate the staging of the delivery trucks?
- Has a well-compacted level gravel footing been installed to support the precast panels?
- Have the concrete panels reached sufficient strength?
- Will a crane be available when it is time to place the panels?
- Are all the materials needed to brace the walls at the jobsite?
- Is sufficient caulk available to seal all joints?
- Are the corners square?
- Is each panel properly aligned before the next is placed?
- Is the first-floor framing in place and are the walls parallel to the floor joists braced inward a minimum of 4 feet before back filling?

10

Constructing Slab Foundations

A level concrete floor is called a *slab.* Slabs can be cast as part of a footing and stemwall, pier-and-grade beam, or basement foundation system. Slabs produced in a single cast with an integrated foundation system are called *monolithic* slabs. Monolithic slabs generally are designed with a thickened edge, which acts as the footing and helps transfer the weight of the house to the ground. This thickened edged may or may not be reinforced with rebar. In special cases, the entire slab is reinforced with rebar or tensioned cables. These reinforced slabs are called *structural* slabs.

Slab foundations have some advantages over other types of foundation systems. They can be produced economically and relatively quickly. In the case of a monolithic slab it is possible to set the forms, install the subslab plumbing, and cast the slab in one or two days. This allows the framing of the house to start on the third day.

In regions with stable soil conditions and shallow frost lines, monolithic slabs can provide an economical foundation system. To meet most codes the top of the slab must be a minimum of 8 inches above the finished grade and the bottom edge of the integrated footing must extend either 12 inches below grade or below the frost line, whichever depth is greater.

While the slab's edge acts as a footing to support the frame, monolithic slabs also must incorporate support under interior load-bearing partition walls. Increasing the thickness of the slab under load-bearing walls usually satisfies this requirement.

Modified monolithic slabs, such as post-tensioned slabs, provide solutions for regions with expansive soils. These slabs also can be built in areas that have a high water table provided the soil is stable.

Casting Monolithic Slabs

The first step in building a cast monolithic slab is the construction of the pad. Remove the layer of topsoil to reduce the possibility that decaying organic materials will leave unwanted voids in the supporting substrata or act as a food source for insects such as termites. Level an area of undisturbed or compacted earth that is 4 to 5 feet larger than the footprint of the home on all sides. This clear space will make it easier to get a concrete truck to all parts of the slab.

Incorporate into the form an integral footing that is 12 inches wide for one-story structures, 15 inches wide for two-story structures, and 18 inches wide for three-story structures. If the footing and slab are cast simultaneously, stake off and dig the footing. In most cases machine digging will provide the best and most economical trench. The exterior edge of the footing must coincide with the exterior perimeter of the structure.

After the footing is trenched, place the forms along the exterior edge of the slab. This makes it is easy to stake and secure the forms. Drive the stakes into the ground. Use diagonal braces, from the top of the form to the ground, to keep the top of the form from pushing out when the concrete is placed.

The top of the form should be level and at least 8 inches above the finished grade. If it is not, use a chalk line to mark the height of the slab on the inside of the form. To make it easier to see this line while the concrete is being placed, drive nails partially into the forms along this line to help show the top of the slab.

Incorporating a Brick Ledge

If a masonry facade will be used cast a brick ledge into the slab by nailing a 2×4 along the top interior edge of the form. This creates a ledge that will support the masonry facade and prevent the moisture that condenses in the void between the veneer and the exterior wall sheathing from seeping under the sill plate and into the residence. When the slab is cast, vibrate or rod the concrete to ensure that voids are not left under the board that forms the brick ledge. A monolithic form with a brick ledge is shown in Figure 10-1.

If the footing is cast separately, the footing also can serve as a brick ledge. Mark the perimeter of the residence on the top of the footing and align the forms for the slab along the center of the footing. If the footing prevents the stakes from being driven into the ground, nail blocking into the footing with concrete nails and run a diagonal brace from the ground to the top of the form. Or connect the forms to outriggers—horizontal 2×s attached to stakes. Then attach an additional stake, referred to as a *kicker*, between the top of the form and the ground to keep the top of the form from pushing outward when the concrete is placed.

A third option is to cast a stemwall using the method described in Chapter 7. This can simplify the forming process, especially when a brick ledge is needed and a 2×12 is not tall enough for a form.

Installing Mechanical Systems

Once the forms are ready or the stemwall is cast, place the subslab mechanical systems. The piping for the wastewater system usually is placed first, followed by the hot and cold water supply lines and then the heating, ventilation, and air conditioning (HVAC) ducts.

Place the mechanical and plumbing systems deep enough that they can be covered by at least 1 inch of fill sand plus the 4-inch slab. Extend the plumbing lines above the finished floor level a minimum of 16 inches and firmly secure them by driving steel pins into the ground so that they stay in place while the slab is cast.

Isolate the water lines from the slab by covering them with a foam sleeve at the point where they run through the slab. This prevents a bond from forming between the slab and the pipes that could create

FIGURE 10-1. Incorporate a brick ledge in the form design if the exterior of the slab will be covered with a masonry facade. Make sure inspections are in before proceeding.

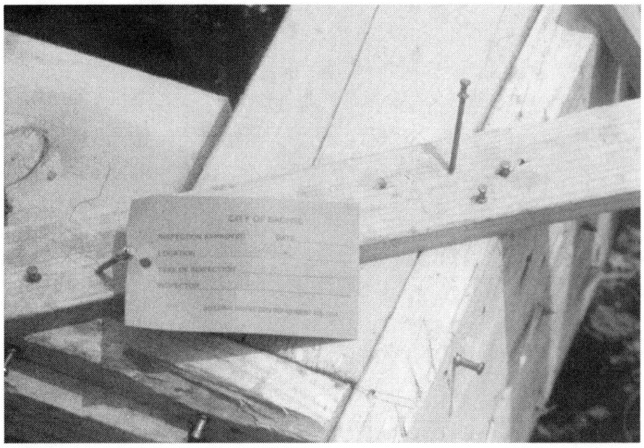

stress on the pipes and eventually cause a leak. It also prevents caustic agents in the concrete from causing the pipes to deteriorate. To avoid the potential for leaks from an improperly sweated joint, do not splice the domestic water line below the slab.

After the domestic water line is placed and before the concrete is cast test the water lines. Attach an air pressure gauge with a tire valve to one line and seal the end of all other lines. Pump air into the lines until the pressure in the pipes reaches 20 psi. Then check the gauge 20 to 30 minutes later. Any decrease in pressure indicates either a defective solder joint or pipe.

Placing the Subgrade

After the mechanical systems are installed and inspected fill the area around the pipes and ducts. Commonly, gravel or sand is used for the subgrade. Mechanically compact the fill using a gasoline-powered tamper with a 4- to 6-square-foot plate that moves rapidly up and down. This eliminates the potential for subsidence after the slab is cast. Spread and compact the fill to allow the casting of at least a 4-inch slab. A subgrade being compacted is shown in Figure 10-2.

Be careful not to dislodge plumbing pipes while placing and compacting the fill. If the pipes are moved, a jackhammer may have to be used on the slab to relocate the pipes or the framing plan may have to be altered. Either solution involves additional costs. Most plumbers leave a pressure gauge on the lines until they return for the rough in. If a gauge is there, have workers check the readings before and after compaction to tell if any accidental damage occurred.

FIGURE 10-2. Compact the subgrade with a gasoline-powered tamper.

After the fill is compacted set intermediate grade stakes if needed. Drive grade stakes into the fill about 10 feet apart to correspond with the top of the forms and indicate the finished floor level of the slab. This helps workers keep the surface of the slab level when screeding. Remove the stakes after the grade is established.

Additional procedures may need to be followed before the concrete for the slab can be placed. If local requirements or site conditions mandate that the edge of the footing be reinforced, place rebar horizontally along the interior of the slab. Typically, two 1/2-inch #4 pieces of rebar are placed 2 inches above the bottom of the trench. If several bars are required, place them as required by the local code. Secure the rebar by placing it on chairs that rest on the ground.

Insulating the Slab

Some localities require that slab foundation systems be insulated. A monolithic slab can be insulated by placing a sheet of polystyrene along the interior side of the form. The insulation is left in place after the slab is cast and a protective wearing surface, such as a coating of stucco, is placed over the polystyrene sheet. Decorative panels embossed with textures that make them look like stone, brick, or other patterns also can be used for this purpose. Insulating panels also are available. These panels are best applied after the slab is cast.

If the footing and slab are cast simultaneously, place sheets of polystyrene foam along the interior side of the footing to add insulation. The foam sheet should almost reach the surface of the slab. Make sure workers do not disturb the sheets while the concrete is placed and the backfill does not push them out of place.

If a stemwall is used, sheets of polystyrene foam can be placed along the inside of the stemwall on the surface of the fill or between the stemwall and the backfill so that they are held in place by the backfill. If the first method is used, break a 4-inch-wide piece off the 24-inch-wide sheets and turn it up to form a break between the stemwall and the slab. Polystyrene foam insulation is shown in Figure 10-3. Also illustrated is how to wrap copper tubing to prevent concrete from coming in contact with the pipes. The subgrade plumbing for the bathtub is in a box filled with sand, eliminating the need for the plumber to break the slab to locate the waste line.

Installing a Vapor Barrier

To keep moisture that has migrated through the fill from infiltrating the concrete slab, some local codes require a vapor barrier. Place a 6-mil polyethylene sheet on top of the compacted fill, taking care to prevent any punctures or tears. If several sheets of polyethylene are needed to provide full coverage, overlap the sheets a minimum of 12 inches. While most regions do not require that the overlapping sheets be cemented together, it is wise to double check local codes to make sure no such requirements apply.

Controlling Cracks

Cracks can be controlled by placing welded wire fabric (WWF) or mesh in the slabs. The fabric's metal wires help hold the pieces of concrete together if it does crack, keeping the concrete surface smooth. The WWF is not considered structural reinforcing material. Some building codes do not require the use of WWF in slabs and many builders do not use it.

If the WWF is used, it should be a 6-inch wire mesh with yield strength of 65,000 psi. Place the fabric on chairs to hold it in the center of the slab. If the slab is thicker than 4 inches, place the mesh 2 inches below the surface. Do not use objects such as broken brick to hold the fabric in position because the concrete will tend to flow unevenly around the wire and decrease the slab's strength.

Using WWF does not eliminate the need for control joints. A 4-inch-thick slab with WWF must have control joints every 36 feet. The easiest way to place a control joint is to wait until the concrete has reached an initial set and then use concrete saws to cut the joint. Saws are available that allow a worker to cut control joints without standing on the slab. These lightweight saws cut easily through the fresh concrete to the correct depth of 1 inch for a 4-inch slab or one-quarter the thickness of the slab. If WWF is not used to control cracks, the joints must be a maximum of 30 feet apart. If the slab is not rectangular, place a joint to separate the offset portion of the slab from the main rectangle.

Polyester, polypropylene, nylon, or fiberglass strands also can be placed in the concrete to minimize shrinkage cracks. When the load is batched, the ready-mix company adds approximately 1 1/2 pounds of fibers in self-dissolving bags that contain millions of interconnected strands for each yard of concrete. This disperses the fibers uniformly throughout the mix. The concrete bonds to the strands and, because concrete shrinks as it cures, shrinkage cracking on the surface is minimized. A bag of fiber mesh is shown in Figure 10-4.

The inert fibers do not corrode or react with any chemicals in the concrete. They also do not affect traditional finishing methods. Although the addition of these filaments increases the per-yard cost of the concrete, the cost should be offset by the elimination of WWF and the labor required to place it.

FIGURE 10-3. Sheets of polystyrene foam placed along the stemwalls help meet code requirements for insulation.

Casting the Slab

The easiest way to cast a slab is to have the concrete delivered to the site in ready-mix trucks. A yard of concrete will fill a 3×3×3-foot area. Most trucks carry 8 to 10 cubic yards of concrete. When placing a 4-inch-thick slab, 1 yard of concrete will cover 80 square feet. Use concrete with a minimum strength of 2,500 psi to cast the slab. Concrete placed at a 6-inch slump generally is best for slabs because it is workable enough to be placed but does not have too much water in it.

FIGURE 10-4. Adding bags of fiber mesh to the concrete mix can reduce surface shrinkage and cracking.

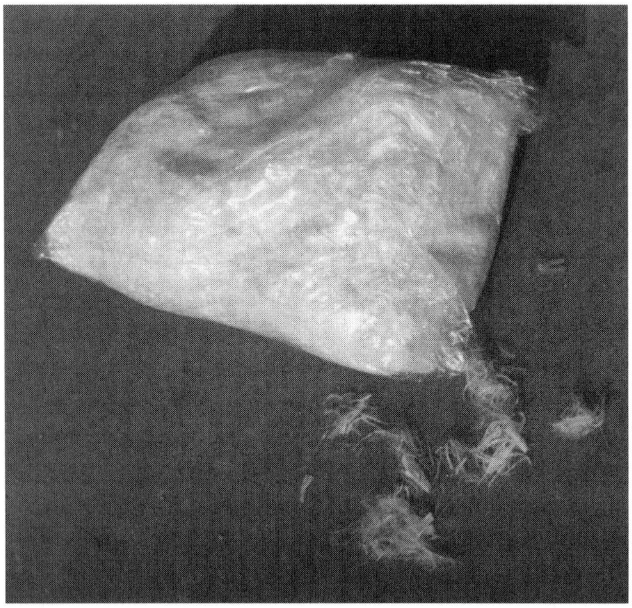

Make sure there is plenty of help when the slab is cast. One person should be in charge of the placing of the concrete. That person should work the chute and give directions to the driver. The chute operator should be able to move the chute and disperse the concrete quickly, but still under control. One or two other workers should even out the concrete with *mud rakes* or *come-alongs*, tools with a wide blade similar to a hoe on a wooden handle. These workers should try to keep the concrete at the desired finished level. During winter months, start the placement early so that the finishers are not trying to complete the job in light provided by the truck's headlights.

As soon as the concrete is placed, screed or level it. To screed the slab, saw a long board back and forth on the surface of the concrete. Use the top of the forms and the reference stakes as control points for the top of the slab. Keep a pile of concrete, 1 to 2 inches tall, in front of the screed. A worker with a come-along can pull away excess concrete or push more in front of the screed as it moves down the slab. Workers screeding concrete and using a come-along are shown in Figure 10-5.

Finishing the Slab

Once the concrete is screeded, bull float or darby it. This procedure creates a smooth finish and sets the large aggregate below the surface. A *darby* is a large float that is used to start the finishing process. A bull float, shown in Figure 10-6, has a handle to which sections can be added. This gives workers the ability to reach areas beyond those accessible with a darby. Some finishers do the initial floating after one truck leaves and before another is ready to unload. Regardless of the timing, the initial leveling and smoothing of the surface must be completed before any bleed water is visible on the surface.

After the slab is bull floated, float it again by hand with a wooden or magnesium float or with a power trowel with float paddles. Floating removes imperfections and compacts the concrete. Too much floating, however, can push the large aggregates too far into the slab and leave just water and *fines*, or cement paste and sand, at the surface. Skilled workers knowledgeable about concrete are necessary to end up with a good job.

Trowel the slab to a smooth hard surface after all bleed water has left the surface. The surface also should have hardened enough to support the workers. Workers on 2×2-foot kneeboards of 3/4-inch plywood can use steel trowels or a power trowel with appropriate paddles to put the smooth, hard finish surface on the slab. The trowels should ring as they

FIGURE 10-5. Level the concrete by screeding, or sawing a long board back and forth on the surface of the concrete.

FIGURE 10-6. Smooth the surface of the concrete with a bull float before bleed water is visible.

are worked across the surface. A power trowel is shown in Figure 10-7.

The final step is to cure the slab. This is one of the most important, but often neglected steps. Because water is essential for the complete hydration of the cement in the concrete, follow procedures to keep too much water from evaporating from the slab. Spray curing agents over the slab as shown in Figure 10-8 or pond water on the slab to cure the concrete. A third option is to cover the slab with sheets of polyethylene to help keep in moisture. Polyethylene sheets used for curing can cause discoloration of the slab's surface. Because the sheets usually do not lay flat on the surface of the slab, the wrinkles change the concentration of moisture. This is cosmetic only and has no effect on the curing process or the final strength of the concrete.

Engineering Structural Slabs

Structural slabs are recommended in regions of the country with very expansive soils. A structural slab has the ability to float on the surface of the ground, rising and falling as the soil expands and contracts. Because the structural slab is cast as a single unit, external pressures usually do not compromise its integrity.

Structural slabs basically are modified monolithic slabs. The most common types of structural slabs are *reinforced slabs* and *post-tensioned slabs*. They are formed following the same basic procedures used for monolithic slabs.

A structural engineer should design reinforced and post-tensioned slabs. The design should consider both the site conditions and the layout of the home's rooms. The number of reinforcing bars used, their spacing, and the diameter of the rebar is determined by the structural engineer to accommodate the specific site and design requirements. The plumbing and mechanical systems also must fit both the floor plan and the design of the structural system. Because of the design work involved, structural slabs generally are more expensive than monolithic slabs.

Casting Reinforced Slabs

Unstable soils, such as expansive clays, can exert extreme forces on residential slab foundations. Reinforcing steel cast into 5- to 7-inch-thick slabs resists the flexural or bending forces exerted by these soils. The reinforcing steel does not start to work, however, until the slab starts to bend or crack. Applying stress to a reinforced slab makes the slab stronger.

Before the concrete is cast, center the rebar vertically within the form by placing the rods on specially designed chairs. Do not use bits of broken brick or

FIGURE 10-7. Use a power trowel to create a smooth hard surface after all bleed water evaporates.

Source: Residential Concrete, *3rd ed., Home Builder Press, National Association of Home Builders, Washington, D.C.; 1999.*

FIGURE 10-8. Spraying a curing agent on the concrete keeps in the moisture required to complete the curing process.

Source: Aluma Enterprises Inc., Toronto, Ontario, Canada.

concrete to support the rebar. At a minimum, the rebar should be 1/2 inch in diameter. Place the steel in two courses, one running lengthwise and the other crossing the width of the foundation to form a mat. Use wire to fasten the bars together where they intersect. Do not weld the intersections because this weakens the rebar. If extensions are needed to accommodate the full length or width of the slab, overlap the extensions with the first set of bars by at least 2 feet and hold them tightly together with wire. The spacing of the reinforcing steel should not interfere with or be compromised by the plumbing and mechanical systems. An additional run of reinforcing steel may be required to bridge around uneven spacing created by obstructions.

If needed, concrete beams can be incorporated into the slab's design to provide additional support for interior load-bearing partitions. Cast with the slab, these beams are created when concrete fills in the depressions left in the fill. The beams usually are 16 to 24 inches wide and may contain additional reinforcement. Make the depressions for these beams, sometimes referred to as thickened slabs, before placing any reinforcing steel.

Use concrete with a minimum strength of 2,500 psi for reinforced slabs. A higher strength may be specified for some structural slabs. The mix should not include an accelerator such as calcium chloride because the additive can promote *oxidation* or rusting of the steel rebar and weaken the concrete. When placing the concrete, take care to avoid disturbing the placement of the rebar. Workers must be careful not to trip on the steel. After placement, cure and finish the concrete as normal.

Installing Post-Tensioned Slabs

Post-tensioned foundation systems build tension into the slab, unlike reinforced slabs that depend on soil conditions to create beneficial pressure. Steel cables with high tensile strength are cast into the post-tensioned slab causing pressure to be applied while the slab is being cast. The tension in the cables compresses the concrete slab, creating a system that resists the strain exerted on it by external forces such as expansive soils. If the soil around the perimeter of the residence expands and lifts, for example, the slab is strong enough to bridge any voids created under it. The cables in a post-tensioned slab are shown in Figure 10-9.

Post-tensioned technology has been used in residential construction since the late 1960s when several projects in Texas and Louisiana used this system to reduce soil-caused problems. Today this system is used nationwide to enable building on difficult soils.

The post-tensioned system should be designed by an engineer to meet the specific site conditions. To assure a quality project, only certified materials should

FIGURE 10-9. Cables, shown here on chairs that hold them above the base of the slab, are placed in post-tensioned slabs before the concrete is placed.

be used and competent workers employed. The Post Tension Institute, located at 1717 West Northern Ave., Phoenix, AZ 85021 (602-870-7540), can furnish information on engineers and contractors familiar with this system.

Start the installation by winding approximately six small cables around a central core to form a 1/2-inch-diameter cable with a *tensile strength* of approximately 20 tons. The steel cables, coated with lubricants and placed in a plastic sheath, are referred to as *tendons*. The tendons are placed using procedures similar to those used to place rebar. Most foundation designs, however, require fewer tendons compared to runs of traditional rebar.

Following the engineered plans, dig trenches for the beams and set, brace, align, and secure the forms. Extra bracing may be required to hold the forms and the system anchors in position while the concrete is placed. Before placing the tendons, make sure the mechanical and plumbing systems are inspected, the fill is treated to prevent insect infestation, and the vapor barrier is installed.

Handle the tendons carefully. A tendon should not be used if its sheathing is ripped. Usually two different tendons are used. One is used in the beams and another is used to reinforce the slab. The tendons are marked to indicate their use.

To hold a tendon above the fill, rest it on chairs placed on the compacted fill about 4 feet on center. Place the tendons per the engineer's recommendations, but do not place them in the center of the concrete vertically or closer than 12 inches from any corner of the house. The tendons should not drape more than 1/2 inch from vertical. Drooping cables will straighten when tension is applied and place unwanted strain on the slab.

Permanently attach one end of each tendon to an anchor fixed to the form, 6 inches down from the surface of the slab. Nail the anchors to the forms to hold them in place until the slab is cast. The *dead-end* anchored portion of the tendon will be cast permanently into the slab. The other end of the cable passes through, but is not secured to, another anchor, called the *live-end* anchor. This anchor is attached to the form on the opposite side of the slab. Remove the sheathing on the cable that passes through the live-end anchor. Pass about 2 feet of cable through the second anchor and the hole in the form that holds this anchor in place. A post-tensioned anchor is shown in Figure 10-10.

When all the cables are securely in place, cast the slab using at least a 2,500 psi concrete mix. Admixtures such as calcium chloride that can corrode the tendons should not be used. After the concrete has cured, remove the forms and leave the anchors and tendons cast into the slab.

Make a cylinder test for each batch of concrete used to cast the slab. Because tension-bearing cables will exert pressure on the slab the concrete must reach a specific strength. Failure to verify the concrete's strength with cylinder tests could result in a dangerous and costly system failure.

After the concrete has cured to 2,000 psi, usually in about 10 days and after the strength has been verified by breaking the test cylinders, tension is applied to the cables. A hydraulic device, called a *ram*, is attached to the cable and braced against the slab. The device essentially pushes against the slab while pulling the cable at gradually increasing levels of force until a force of 28,000 pounds is reached. This procedure also stretches the cable. When properly tensioned the distance that each cable is stretched should equal .0833 inches per foot of length.

Next, wedges, called *dogs*, are inserted between the tendon that is precast in the slab and the live-end anchor. When the pressure on the tendon is released, it shrinks slightly against the wedge and locks itself to the anchor. Although a slight amount of tension is lost in the process, a significant amount of tension remains in the cable. This tension adds to the slab's ability to withstand the extreme forces exerted by expansive soils.

FIGURE 10-10. The cables are passed through anchors positioned at each end.

Chapter Checklist

When working with slab foundations consider these questions.

- Have topsoil and other organic material been removed from under the proposed slab?
- Has the site been leveled to allow for the placement of forms?
- Are the sides of all trenches for beams dug straight and the bottoms leveled? Has all loose dirt been removed from the trenches?
- Are all the forms leveled and braced? Are the tops of the forms 8 inches above the finished grade and the bottoms of the beams 12 inches below the finished grade?
- Have mechanical systems been installed and inspected? Has all plumbing passing through the slab been placed in sleeves?
- Has the fill been placed and compacted to provide the required slab thickness?
- Are all pockets for thickened slabs properly placed?
- Has the fill been treated for insects?
- Is the vapor barrier in place?
- Is reinforcing steel in place and securely tied?
- Is the WWF properly installed?
- Is all insulation in place?
- Are all necessary inspections complete?

When working on a post-tensioned slab, add the following to the checklist.

- Are tendons spaced per plans?
- Are anchors secured to the form 4 inches below the top of slab and 12 inches from the corners?
- Do the tendons drape between chairs? They should not.
- Are the tendons secured to the dead-end anchor and passed through to the live-end anchor?
- Is the equipment needed to make test cylinders on hand?

Before applying tension to the tendons, be sure that:

- The test cylinders broke at the proper psi.
- The hydraulic ram is calibrated properly.
- The tendons are stretched to the proper length.

When placing the concrete for the slab, remember these tips.

- Schedule concrete delivery so that the project can be finished before dark.
- Use concrete with no more than a 6-inch slump and verify that the psi meets specifications.
- Have enough help to place and finish the concrete properly.
- Do not over-finish the concrete.
- Do not trowel until all bleed water is gone.
- Use appropriate admixtures to help cast the slab.

11

Building with Masonry Systems

For more than 100 years concrete masonry units (CMUs), more commonly called concrete blocks, have been used to construct residential foundation systems. CMUs hold a strong market share in residential foundation construction because they work well. They can be used to build most basement walls. When conditions require reinforced walls the cavities in the blocks provide a place in which reinforcing steel can be installed.

Most CMUs are manufactured under controlled conditions that ensure the consistency of the raw materials and the production process so that the blocks perform consistently in the field. CMUs also are tested to assure minimum compliance with required strengths and dimensional tolerances.

CMUs are available in two grades. Grade S blocks are for use in load-bearing walls and may be exposed to weather or moisture. Grade N blocks are limited to above-grade use where they are protected from the weather. Only Grade S blocks should be used in residential foundations.

Depending on the structure of the foundation, its exterior facing, and the height of the backfill, building codes set minimum requirements for the thickness of the blocks and the placement of steel reinforcements, if any. Local code officials may modify basic code requirements as needed to accommodate site conditions. The table in Figure 11-1 provides the minimum thickness of masonry walls with backfill. Figure 5-2 in Chapter 5 provides wall construction specifications by seismic zone.

Two main factors determine whether or not concrete blocks are suitable for a project. The first is availability of the product. Most blocks are used within a 100-mile radius of where they are produced because of shipping costs. The second factor is the availability of skilled labor. It may be difficult in some parts of the country to find a crew skilled enough to do a quality job, especially because the average age for masons is more than 50 years. In some areas masons prefer to work with bricks rather than

FIGURE 11-1. The thickness of the walls required for the foundation is based on site conditions.

FOUNDATION WALL CONSTRUCTION	NOMINAL THICKNESS[3] (inches)	MAXIMUM DEPTH OF UNBALANCED FILL[1] (feet)
Masonry of Hollow Units, Ungrouted	8 10 12	4 5 6
Masonry of Solid Units	6 8 10 12	3 5 6 7
Masonry of Hollow or Solid Units, Fully Grouted	8 10 12	7 8 8
Plain Concrete	6[4] 8 10 12	6 7 8 8
Rubble Stone Masonry	16	8
Masonry of hollow units reinforced vertically with No. 4 bars and grout at 24 inches on center. Bars located not less than $4^1/_2$ inches from pressure side of wall.	8	7

Source: Council of American Building Officials (95) Table 404.1.1a.

blocks. Before planning a CMU foundation be sure a mason is available and that the price to lay the blocks fits the budget.

Most manufacturers deliver blocks in *cubes.* Each cube holds 108 blocks and typically includes one corner block for every three stretchers. Most suppliers will split the cubes if a specific number of blocks are ordered. Sash and half blocks must be ordered separately.

Selecting Blocks

Grade S concrete blocks are manufactured in a wide variety of shapes, sizes, weights, and colors, all of which are suitable for foundation construction. Builders have the option of using one type of block or mixing several types to produce a foundation that is both structurally sound and aesthetically pleasing.

Determining Block Size

Most blocks are manufactured in *modular* sizes. Blocks used to build residential foundations generally are 16 inches long and 8 inches tall in *nominal* dimensions. The actual length and height of the block, however, are less than the nominal length and height. The difference between the actual and nominal measurements, usually about 3/8 of an inch, is built into the block's design to accommodate the thickness of the mortar joints that hold the blocks together. Some manufacturers produce blocks that vary in height. These shorter blocks can be used to compensate for footings that are not level.

Blocks typically come in widths of 4, 6, 8, 10, or 12 inches. These dimensions also are nominal, with actual dimensions reduced 3/8 of an inch. Because there is no need to adjust for the thickness of mortar joints, height measurements typically reflect the actual size rather than a nominal size. *Half blocks,* which are nominally 8 inches long, also are available as a standard product.

Most block styles fall within three main types. *Stretcher* blocks are used to construct the main body of the wall. They have two flat-face surfaces and indented ends. Half blocks are like stretcher blocks, but half the size. *Corner* blocks have two flat-face surfaces and at least one flat-end surface. *Sash* blocks have two flat-face surfaces and one flat-end surface with a vertical groove that will accommodate the side sash of a window or doorframe. A variety of blocks are shown in Figure 11-2.

Corner blocks are produced in 10- and 12-inch sizes with one 8-inch side to allow the mason to keep the end joints of the blocks on 8-inch increments. The shape of *trough* and *beam* blocks, when laid in a course in a wall, form a hollow, horizontal channel into which steel rebar and grout can be placed. This enables workers to construct a job-built beam into the wall at the jobsite. Trough and beam blocks are shown in Figure 11-3.

FIGURE 11-2. Masonry blocks are available in a variety of shapes as shown in this catalog page.

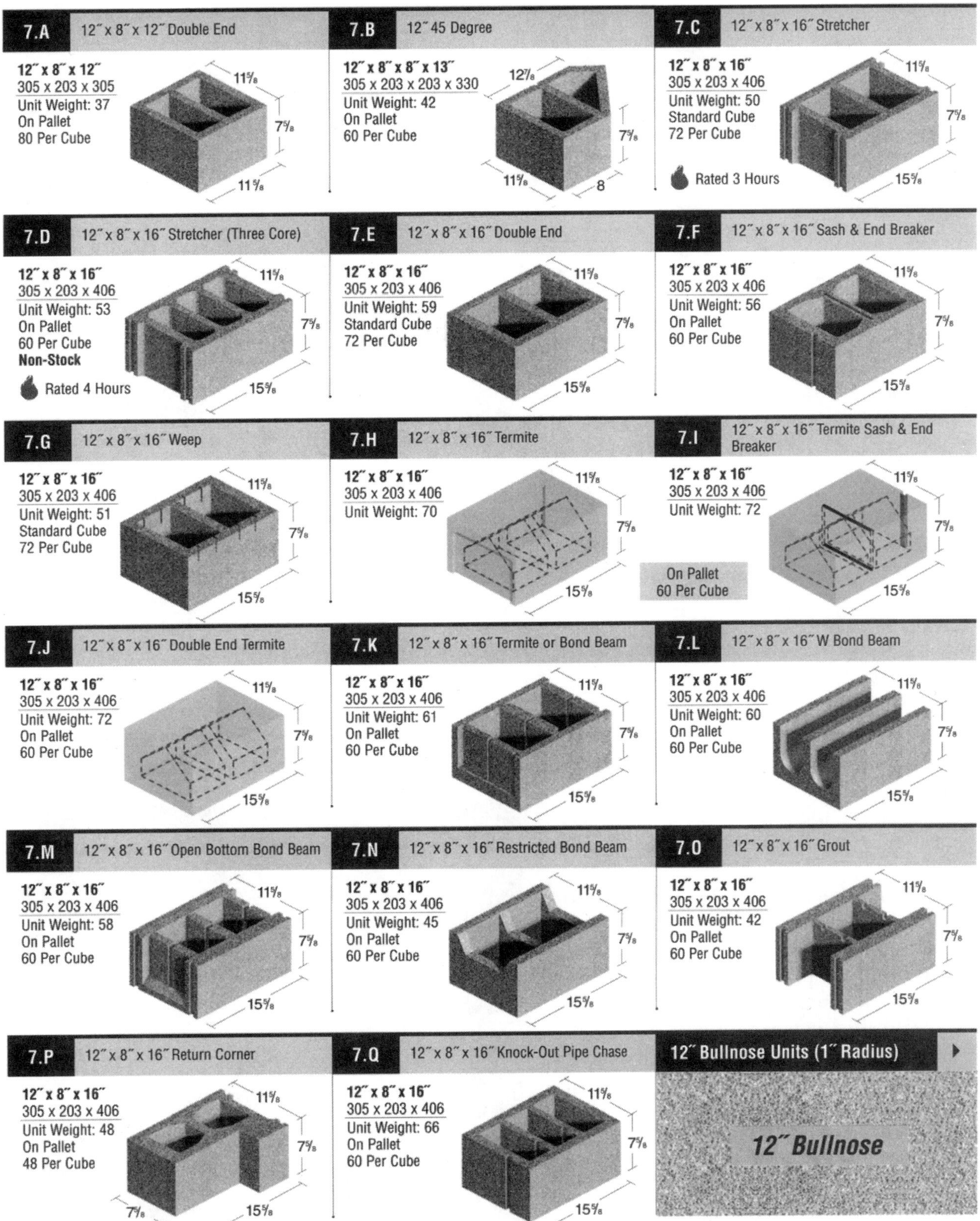

Source: Oberfield's Inc., Delaware, OH.

FIGURE 11-3. Trough and beam blocks form a hollow, horizontal channel into which steel rebar and grout can be placed.

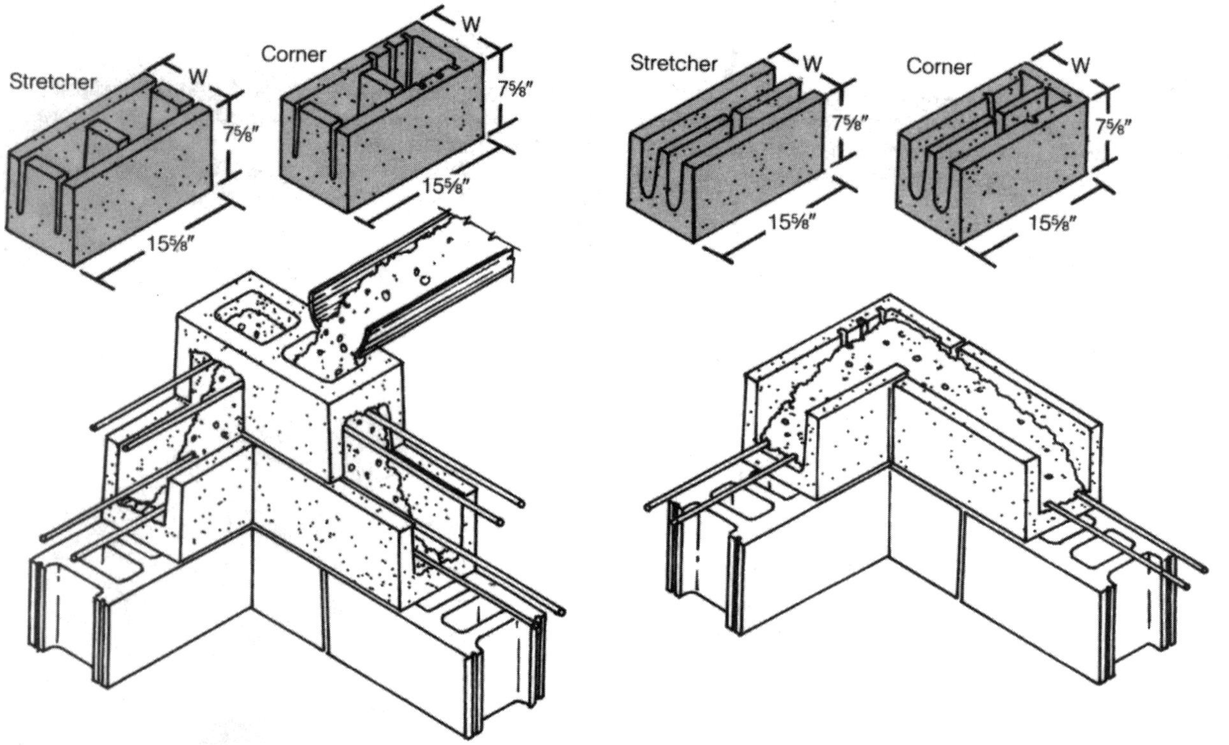

Source: Besser Company, Alpena, MI.

Lintels or *header* blocks also are available for use above doors and windows. These reinforced blocks, designed to span openings, are available in 16-inch increments and have jointed surfaces that mimic the look of smaller individual blocks. Header blocks are cast with steel reinforcing dowels in them to give them a greater load-bearing capacity than normal blocks.

Blocks also are available with foam inserts to increase their insulation value. Some manufacturers produce a block with a thermal break to obtain a greater R-value in the completed wall. The completed cost of a foundation constructed with these blocks generally is competitive with a traditional block wall insulated to the same R-value when all material and labor costs are included. An insulating block is shown in Figure 11-4.

FIGURE 11-4. Blocks with foam inserts increase the insulation value of the wall.

Source: Concrete Block Insulating Systems, Inc., West Brookfield, MA.

Customizing the Appearance of the Block

Most blocks used in residential foundations have two *face surfaces* and a hollow interior. The face surfaces and ends may be created with decorative surfaces. Some manufacturers cast a large unit and then split it into several blocks. The split-faced blocks created in this fashion have a face that resembles stone. An array of textured face blocks is shown in Figure 11-5.

Coloring also can be added to blocks during the manufacturing process. If a specific colored block is requested, make sure that *all* colored blocks used on the project come from the same manufacturing run. Even minor variations in the amount of raw materials can cause color variations, so the best bet is to buy blocks from the same batch.

Reducing Block Weight

The block's hollow interior serves several purposes. It reduces the material required to produce each block and reduces the weight of the block by 30 percent. The hollow interior also allows insulating materials or reinforcing steel and grout to be placed inside the wall.

Different raw materials are used to create blocks of various densities or weights. Blocks that are the same size and have the same strength can vary in weight from 28 to 44 pounds. Lighter weight blocks usually are more expensive. For example, a 28-pound block made with perlite usually sells at a 15-percent premium. The additional cost for the lighter blocks may be worth the reduced labor costs, however, because masons often charge less to lay lighter blocks. If a mason lays 300 blocks in a day, using 28-pound blocks can reduce the total weight of the materials by more than a ton.

FIGURE 11-5. The face surface of decorative masonry blocks resemble stone.

Source: National Concrete Masonry Association, Herndon, VA.

Considering Nontraditional Products

Two types of blocks first introduced in Europe also should be considered. *FASWALL* is a block manufactured using a composite of wood fibers and cement. Autoclaved Aerated Concrete, sometimes referred to as Autoclaved Cellular Concrete, is made from sand, cement, lime, and water with the addition of a chemical that creates tiny air pockets in the block. These products were used successfully in Europe for several decades before coming to the United States.

Both types of block are lighter in weight than traditional CMUs. Because they weigh less they are produced in blocks up to 12×24 inches in size. These blocks also are easier to work with and can be handled as easily as normal block. They can be cut with a saw using a woodcutting blade. They also provide a higher R-value than traditional blocks. Before using either of these products check with local code officials. It may be hard to convince code officials that these are acceptable equivalent materials.

Reinforcing Concrete Block

Site conditions such as the height of the water table, height of the backfill, and the potential for a seismic event determine the type of reinforcement required.

Concrete block can be reinforced using several techniques. *Pilasters*, which provide vertical reinforcement, can be incorporated into the wall during construction. A pilaster is made by turning a block at a right angle so that one of its ends projects into the foundation. Another block then is laid next to it. In the next course two blocks are laid next to each other in the plane of the foundation. This produces a column within the foundation wall. Figure 11-6 shows various methods for constructing a pilaster in a CMU wall.

Pilasters usually are formed on the interior side of the foundation, with the reinforcing rebar also on the tension side, or inside of the structure. This produces a clean line on the exterior and makes it easier to install an exterior damp-proofing or water-

FIGURE 11-6. Pilasters, which provide vertical reinforcement, can be constructed using a variety of methods.

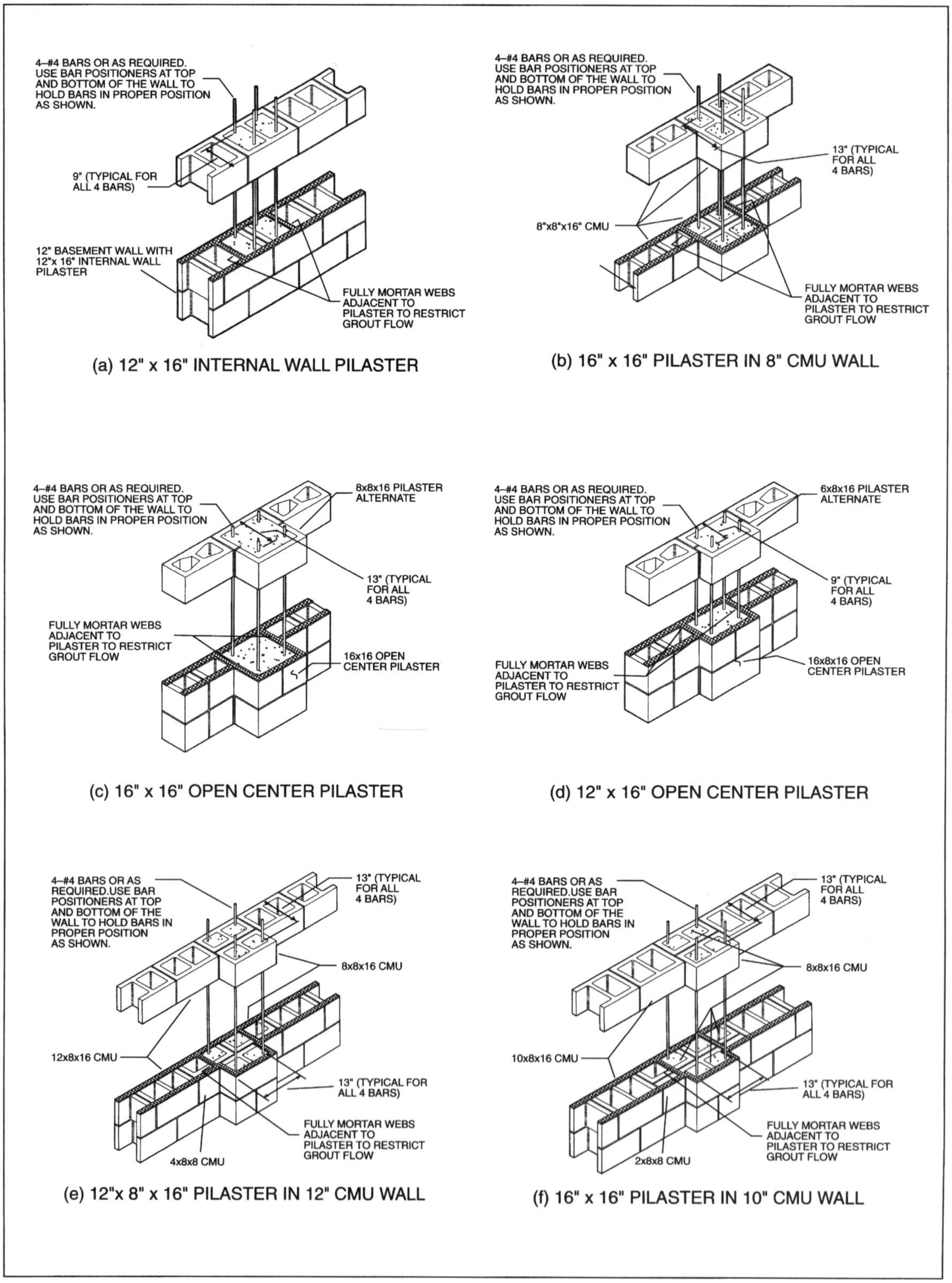

Source: Besser Company, Alpena, MI.

proofing membrane. Because pilasters are on the inside, however, they can alter the way in which interior basement walls are finished.

Another method of reinforcing blocks is to place steel dowels or 2×2-foot pieces of angle iron into the vertical cores, or *cells*, of the blocks and then filling the cores with a cementitious grout, as shown in Figure 11-6. The grout should be a mixture of portland cement and coarse sand. In a third method, the cores of all the blocks in the foundation wall may be filled with grout. A funnel, shown in Figure 11-7, can help insert the grout into the blocks.

Lateral strength can be gained by placing a steel wire truss into the joint between every other course of block or by using specially designed blocks with horizontal reinforcing steel.

Constructing with Blocks

Except for the differences in wall heights, the techniques for constructing slab-on-grade, concrete-block foundations, and crawl spaces are very similar. Cast-concrete footings support concrete-block foundations. The footing should be designed to meet the site's soil conditions and be cast level. If the footing is not level, use blocks that are less than 8 inches tall or cut the blocks to compensate for minor irregularities in the footing.

FIGURE 11-7. A box-funnel simplifies the placement of grout in the voids of the blocks.

Because of the modular size of the blocks, it is helpful if the design places doors and windows at distances that are multiples of 8 inches in from the corners so blocks do not need to be cut. Blocks that are cut to adjust for door and window openings can be placed at any point in the wall.

After the final course is laid, sill anchor devices are inserted into the cores of the block and grouted into place, as shown in Figure 11-8. The anchor devices attach the framing of the structure to the foundation. Some codes require that the anchoring device extend down through the top course of block into the course below. Some builders also use solid blocks or fill the cores of the top course of blocks to create a solid base for the framing.

Incorporating a Drainage System

Some local codes require that the first level be built with blocks specifically designed for the first course.

FIGURE 11-8. Place anchor devices in the top course of blocks according to code specifications.

These special blocks have notches or *weep holes*, shown in Figure 11-9, molded into the bottom. The weep holes allow water to pass from the cores of the blocks into the foundation drainage system. If these blocks are not available masons must cut or chisel the weep holes into regular blocks to meet this code requirement.

As an alternative, special products may be used in place of the weep holes. For example, a special corrugated metal drainage device, designed to be used with regular blocks, channels water away from the footing. Another product provides a channel that directs the water to the footing drain and serves as an isolation joint between the basement floor and the foundation wall. The isolation joint prevents the concrete in the floor from binding to the wall and reduces the stress on the basement floor as it shrinks with age.

If this system is not used an isolation joint still should be placed between the foundation wall and the basement floor to limit shrinkage cracks in the basement floor. This can be done by placing asphalt-impregnated joint material, such as roofing felt or polyethylene film, along the basement wall up to the level of the finished floor and then spraying the area with a releasing agent. A drawing of a worker installing an isolation joint is shown in Figure 11-10.

Some builders install drainage tile in the wall. The drainage tile is placed on the footing and under a beam or trough block. The interior side of the block is notched to allow the tile to pass through the block and connect to a sump pit. Placing the drainage tile in the wall protects it from damage during back filling. Chapter 4 describes methods for creating drainage systems in block walls.

Specifying a Cement

CMUs are laid with masonry cement. The type of cement depends on the specific site conditions. Type *N*, or low-strength, cement is used only when no lateral force will be exerted on the wall. It should not be used in foundations except for light work such as attaching a masonry veneer. Type *S*, a moderate-strength mortar made from a mix of two parts Type II masonry cement and one part portland cement, works well for most residential basements. Type *M* mortar is a high-strength mixture of equal parts Type II masonry cement and portland cement. Type M mortar is used for basements built in difficult soils, such as clays, that can exert extensive lateral force against the foundation wall.

FIGURE 11-9. Weep holes in the first level of blocks allow water to pass from the cores of the blocks into the foundation drainage system.

Some cement manufacturers produce premixed Type S or M masonry cements. These premixed products usually are no more expensive than purchasing the products and mixing the ingredients at the site. Premixed products, however, do reduce mixing time at the jobsite. All cements should be mixed with clean masonry sand and water to make mortar. Once mixed, use the mortar within 2 1/2 hours. Dampen the mortar with water to adjust its consistency during this period. A mortar chart is provided in Figure 11-11.

Bracing the Finished Wall

Until the floor system is framed and the floor sheathing is placed, brace freestanding block walls to prevent any possibility of collapse. A freshly laid 8-foot-tall block wall that is not braced can fail under a 35-mile-per-hour (mph) wind load. Back filling the excavation also can cause the wall to distort or fail and should not be completed until the basement floor is cast and the first-floor framing is completed. A bracing method for block foundation walls is shown in Figure 11-12.

The construction of the first-floor system acts as a diaphragm, increasing the structural integrity of the block foundation walls. Do not remove the bracing until the flooring system, including the bridging between the floor joists, is in place. The bridging prevents the rim joists, which are placed

FIGURE 11-10. Isolation joints limit shrinkage cracks in basement floors.

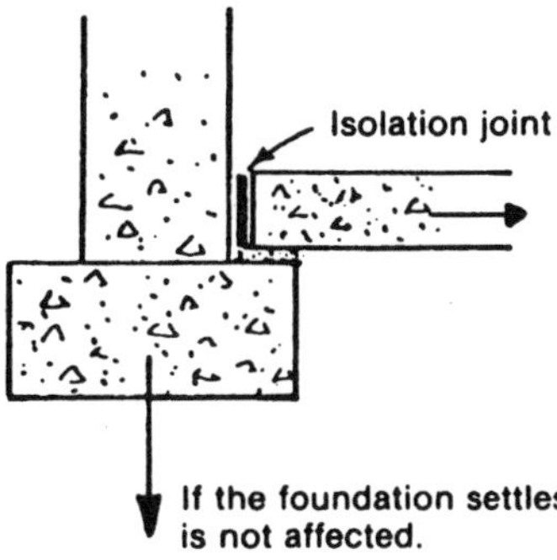

a. Slab-to-wall isolation joint

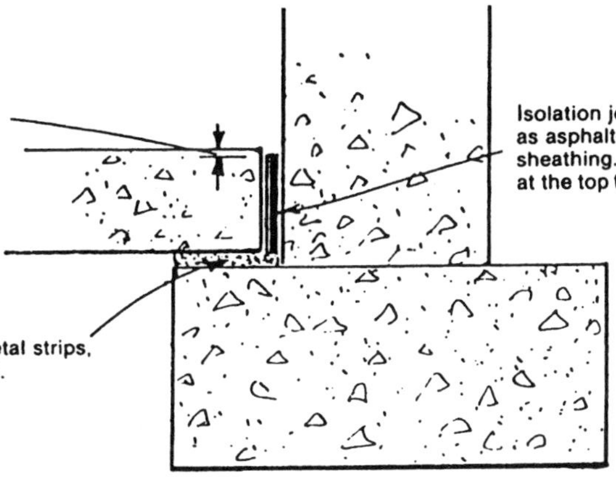

b. Details of slab-to-wall joint

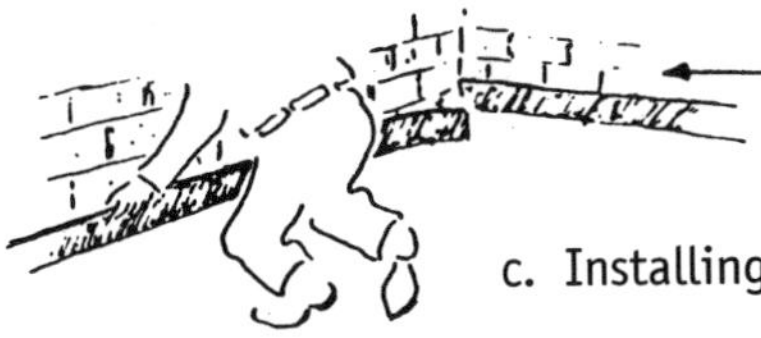

c. Installing isolation joint filler material

Source: Residential Concrete, *3rd ed., Home Builder Press, National Association of Home Builders, Washington, D.C.; 1999.*

FIGURE 11-11. Mixing different ingredients creates mortars of different strengths.

Mortar Type	Parts by Volume: Portland Cement	Masonry Cement	Hydrated Lime	Loose Damp Sand
H	1	—	1/4	Between 2 1/4 and 3 times the sum of the volumes of cements and hydrated Lime
High Strength	1	1	—	
S	1	—	1/4 to 1/2	
Moderate Strength	1/2	1	—	
N	1	—	1/2 to 1 1/4	
Low Strength	—	1	—	

parallel with the foundation wall, from twisting. Because the rim joists hold the top of the wall in place, it is essential that the bridging within the three joists adjacent to the foundation be secured before the backfill is placed.

Engineering Walls to Support Veneers

If the foundation wall will support a veneer of bricks or stone masonry it must be engineered to support the additional load. If the facade starts at the finished grade, use two sizes of block to construct the foundation wall. Lay the larger blocks to an elevation at or just below the finished grade and use blocks 4 inches or less in width to lay the remaining above-grade courses. A brick ledge constructed this

FIGURE 11-12. Brace block foundation walls from the inside to counteract exterior lateral pressures, such as wind.

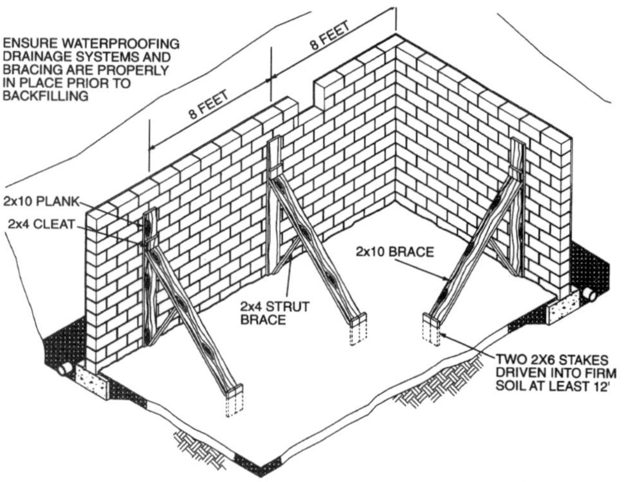

Source: National Concrete Masonry Association, Herndon, VA.

way is shown in Figure 11-13. Note that the angle iron is grouted into the block to form a pilaster.

The change in the width of the blocks allows room for the veneer to be added on the upper courses while providing maximum support from the lower, wider courses. Making the transition below grade hides the blocks from view and leaves only the attractive veneer visible.

Insert corrugated metal wall ties into the joints above the first course of smaller blocks. Use the wall ties to connect the veneer to the structural wall and help hold it in place. Attach the sill for the framing to anchors in the block wall.

Finishing the Above-Grade Wall

Some builders apply a parge coat to the exterior of block walls. This coating, usually either a mixture of one part portland cement to three parts sand or Type M mortar, serves two purposes. Some people find its stucco finish more attractive than the exposed block. The coating also fills the porous surface of the block, helping to protect it against moisture infiltration. With or without a parge coat, masonry block foundations must be protected against water.

To parge a wall, apply a 1/2-inch-thick layer of the cementitious material to a clean, damp wall. If two layers of coating are used, rough up the first layer by rubbing it with a wire brush or a comb with metal tines when it is partially set. After the first coat has been in place 24 hours, dampen the wall and apply the second layer. Dampening the wall fosters the best bond between the two layers. Trowel the exterior layer to a smooth hard surface.

Damp-Proofing the Wall

Cementitious and latex paints, as well as asphalt-based products can be used to damp-proof basements. These products can be applied using a brush, roller, or spray pump. Many products require two coats to be effective. Follow the manufacturer's recommendations for best results. More information on damp-proofing and waterproofing basements is provided in Chapter 4.

Tying Together Walls of Different Heights

Block construction often is used for foundations that have different levels. For example, the level of the garage floor may be higher than that of the main basement floor, or the house may have a partial basement and a crawl space under the remainder of the home. To construct multilevel foundations with minimal excavation and backfill, cast the footings at the different levels. For the easiest installation, design the home so that the distances between the various levels are evenly divisible into 8-inch spaces.

FIGURE 11-13. Use a wider block at the bottom to create a brick ledge in a block foundation wall.

Careful measuring helps compensate for the minor irregularities in the height of the block.

Lay the lower courses of blocks up to the top of the higher footing. Then position a lintel or beam block so that it spans the distance from the footing to the block wall. As shown in Figure 11-14, the beam should span from the footing and rest fully on the lower wall, and the end of the beam should be flush with the interior of the wall. The first course of blocks is laid on the higher footing. The second course of blocks should cross the beam and join the course on the lower wall. Most masons tie the walls together by alternating the courses. Every other course of blocks from the lower wall is laid into the wall for the lower portion of the foundation.

Constructing Surface-Bonded Block Foundations

CMUs also are used to construct foundation walls without mortar. Dry stacking and surface bonding the foundation walls does not require as much skill as traditional block construction. Even though the materials are more expensive, the use of less expensive labor may make this an attractive alternative. The time it usually takes to construct either method is comparable.

Dry-stacked foundation walls are laid out in the same basic manner as mortared block walls. The stacked blocks are held together by a special surface-bonding material, generally made up of a mixture of cement, sand, fiberglass strands, coloring, and water. These products come in premixed bags and are available under brands name such as *Surewall* and *Quickwall.* The surface bonding mixtures can be prepared in masonry mixtures by adding specific amounts of potable water. Adding incorrect amounts of water can weaken the product or make the application more difficult.

Footings under dry-stacked walls must be level because there are no joints in the wall that can be used to compensate for a first course that is not level. Set the first course of blocks in a full bed of mortar. Do not use a head joint; the blocks should touch each other. If regular blocks are used, cut or grind the blocks before they are placed to remove any irregularities that might affect the level or plumb of the wall. Stack the remaining courses of blocks on top of the first course, as shown in Figure 11-15. Check the growing walls to ensure that they are both level and plumb.

After all the blocks are stacked, apply a surface-bonding mixture to the dampened surface of the blocks. Coat both the interior and exterior sides of the walls. A least a 1/8-inch-thick coating generally is required to ensure a good bond. Apply the product with a trowel, or spray it on using the proper equipment. Manipulating the stucco-like bonding material can produce several surface textures. Most surface-bonding materials do provide damp-proofing, but additional products must be used to achieve below-grade waterproofing.

FIGURE 11-14. Position a lintel or beam block so that it spans the distance between stepped footings.

FIGURE 11-15. Once a level course is laid, the remaining blocks can be dry-stacked.

Reinforcing Surface-Bonded Walls

Many manufacturers of surface-bonding cements recommend that basement walls incorporate horizontal and vertical reinforcement. The horizontal reinforcement usually is placed midway up the wall in the fourth course and also in the top course, approximately 8 feet above the footing.

To accomplish this, place trough blocks at these levels. Insert 1/2-inch rebar in the trough blocks and fill the blocks with grout. Then place vertical 1/2-inch rebars 6-feet on the center of the cores of the blocks. After the steel is grouted at the 4-foot level, dry stack the remaining block courses and repeat the process for the top course. Install framing anchors in the top course.

Using Ground Spline Blocks

Because dry-stacking techniques do not use mortar joints masons must measure and cut most CMUs to ensure proper, level stacking and adherence to the foundation dimensions. To eliminate this inconvenience, a special type of dry-stacking block is available from some block manufacturers. Called *ground spline blocks* (GSB), these blocks are produced in full-size molds so that their nominal and actual dimensions are the same.

The top surface of the GSB is machined or ground to eliminate any surface irregularities or imperfections created during the casting and curing processes. A small groove also is cast lengthwise into the middle of both the top and bottom surfaces of the block. During construction, a thin 48-inch-long splint can be inserted into the grooves to help hold the blocks together.

As with conventional masonry block techniques, anchoring devices must be grouted into the top course of GSBs to properly attach the framing to the foundation. Backfill also should not be placed until the framing for the first floor is complete to provide support for the top of the walls.

Building in Extreme Weather

When the temperature rises above 90° F or falls below 40° F special procedures must be followed to ensure the construction of a quality block foundation.

When the temperature exceeds 100° F or the temperature is above 90° F with more than an 8-mph wind, follow hot weather procedures, which include keeping the materials cool. Keep the cement and the water for the mortar in the shade. It is best to keep water in a 55-gallon drum because water in a hose can reach scalding temperatures. Keep the sand damp. Spray water into the mixer to cool it. Dump out the water before mixing each batch of mortar. Dampen the wheelbarrows and mortarboards before using them. Be prepared to work fast. Once the mortar is spread, the block can set within one minute. After the blocks are set, fog the wall with a fine spray of water to keep the mortar from drying too quickly and weakening the bond.

Cold weather brings different problems. If the mortar around freshly laid blocks freezes it will not provide the bond needed to hold the blocks together. Take measures, such as building a tent to shelter the building site, to prevent the wall from freezing. One easy way to tent the site is to set up and brace trusses every 8 feet and then cover them with a sheet of polyethylene film. If it isn't too cold, a kerosene heater might provide enough heat to keep the wall from freezing. Use the heater carefully, however, because the fumes can create an atmosphere harmful to humans.

Do not use frozen materials. If the sand is heated, it should not be contaminated with ashes. Water must remain potable and should be heated to a temperature greater than 40° F before being mixed with the dry cement. The use of admixtures is not advisable because of their corrosion potential, which might reduce the strength of the anchoring devices. The blocks themselves must not be frozen or covered with frost. The best way to deal with cold weather is to have the foundations built before it arrives.

Building Stemwalls

Stemwalls for slab-on-grade foundations or crawl spaces also can be built with blocks. As with all foundation systems the base of the footing must be 12 inches below the finished elevation or below the frost line, whichever depth is greater. Also, the top of the foundation must be 8 inches above the finished grade. Vents must be provided in crawl spaces. Vents sized 8×16 inches are available that fit into the block matrix without disturbing the normal bond pattern of the block.

Chapter Checklist

Before using CMUs to build a foundation, check the following.

- Are the footings level? If multiple footings are used, are the footings spaced proportionately to the size of the block?
- Are all the steps built into the footing designed in 8-inch increments?
- Have beam blocks or cast beams been ordered in lengths sufficient to span over doors and bridge the distance between stepped footings?
- Have the dimensions and the placement of all windows, doors, and utility knockouts been verified?
- What procedures are in place if the blocks will be placed in cold (below 40° F) or hot (above 90° F) weather conditions?
- Is an adequate supply of potable water available for mixing the mortar?
- Is the appropriate mortar available?
- Are wire trusses on hand to provide lateral reinforcement in horizontal joints?
- Does the code require pilasters or grouted reinforced cells?
- Are sufficient quantities of anchoring devices available and placed as needed?
- Does the foundation have adequate drainage and will appropriate methods be used to protect the blocks from water damage?
- Will the walls be properly braced before the excavation is back filled?

12

Designing Pier, Post, and Piling Foundations

If the soil on a site is unstable or not compacted, has a low load-bearing capacity, or otherwise is unable to adequately support a structure, one solution is to use a foundation system that reaches down to the stable soils below. Piers, posts, and pilings are three related foundation systems that can accomplish this. Even on stable soil, these systems often can be placed in a shorter time and more economically than a footing-and-stemwall system. These systems usually are not used for basements or structures that have an interior portion of the foundation below grade.

Generally *piers* are concrete cylinders that are cast into holes drilled at the building site. *Posts* are treated wood members that rest on top of a concrete pad cast at the bottom of a hole. *Pilings* are treated wood, steel, or reinforced concrete members that are driven directly into the earth. Piers and posts support the weight of a house by transferring the load to a layer of stable soil below the surface. Pilings may transfer the load to stable soil or they may use friction to support the weight of the house by balancing the load against the pressure exerted on the piling by the soil into which it has been driven.

The consequences of moisture at the building site must be considered for these foundation systems. Water left to pond near a structure built with one of these foundations can adversely affect the system's integrity. The soil under the residence should slope away from the center at a fall rate of at least 1-inch per foot to prevent water from pooling. An adequate storm runoff design also is required.

Building Pier-and-Grade-Beam Foundations

By reaching below undesirable soils to bear on more stable subsurface soils, piers solidly anchor the foundation even on difficult sites. The grade beams span from pier to pier over the undesirable soils to transfer the weight of the house to the piers and the stable soil below.

Pier-and-grade-beam systems usually are designed to suit the conditions of each site and the configuration of the residence. Soil tests are taken and examined to determine the depth at which stable soils can be found and their bearing capacity. Using this information and the design load of the proposed residence, the precise placement, spacing, and diameter of the piers are calculated, as well as the size of the grade beams and the required reinforcing materials. A pier layout is shown in Figure 12-1.

Pier-and-grade-beam foundations can result in significant savings over other types of foundations. For example, for a simple rectangular 24×40-foot residence a 12-inch-deep by 24-inch-wide footing requires 9 1/2 yards of concrete. On the other hand 12-inch-diameter piers drilled 4 feet deep every 8 feet around the perimeter require less than 4 yards of concrete. Assuming the stemwall is 4 feet high and 8 inches thick, almost 14 yards of concrete could be saved with a pier-and-grade-beam foundation. This comparison does not consider the cost of any reinforcing steel that might be required.

FIGURE 12-1. The pier layout is based on site conditions and the configuration of the residence.

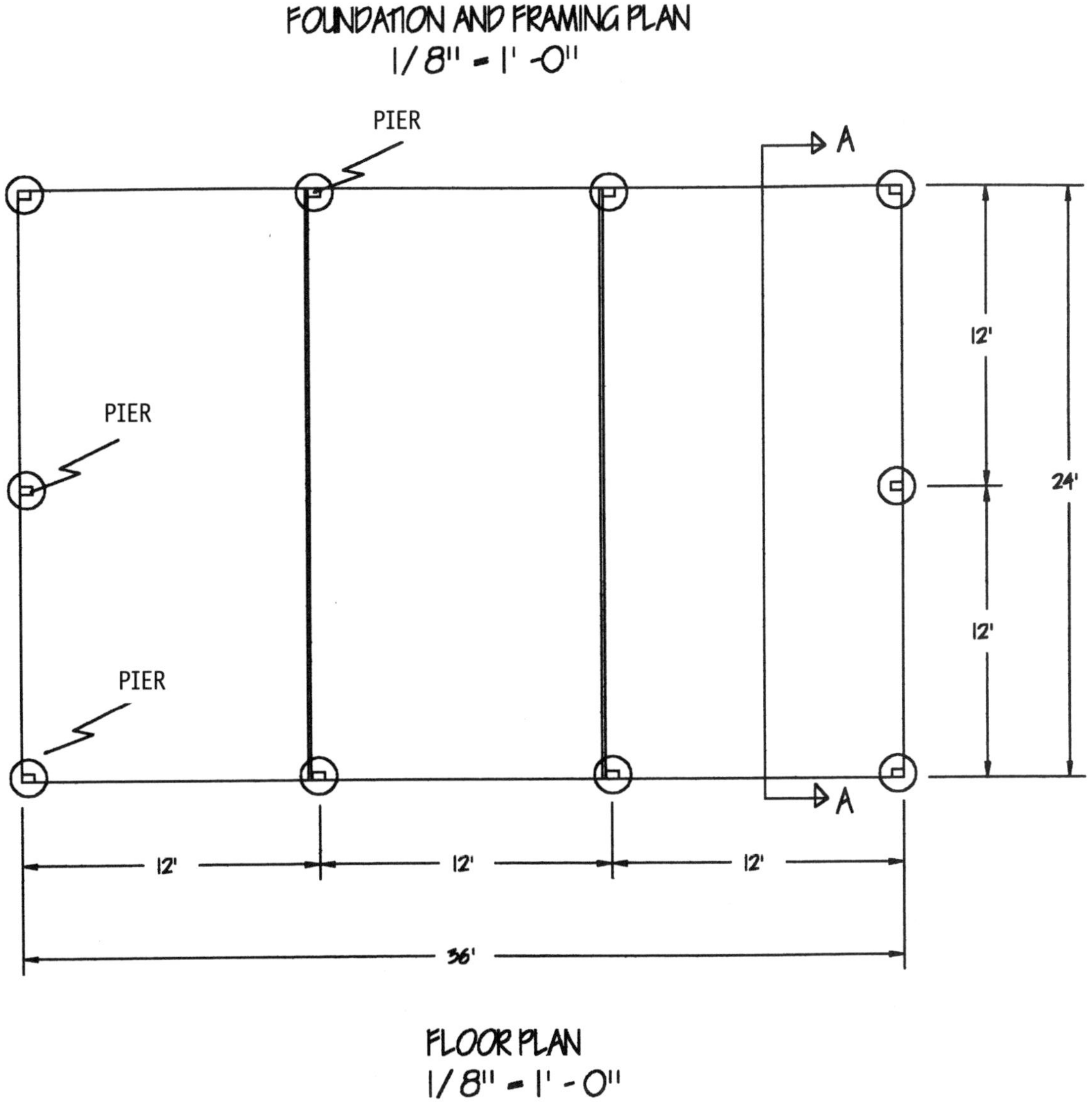

FIGURE 12-2. The pier section illustrates the size, depth, and placement of the pier through the soil strata.

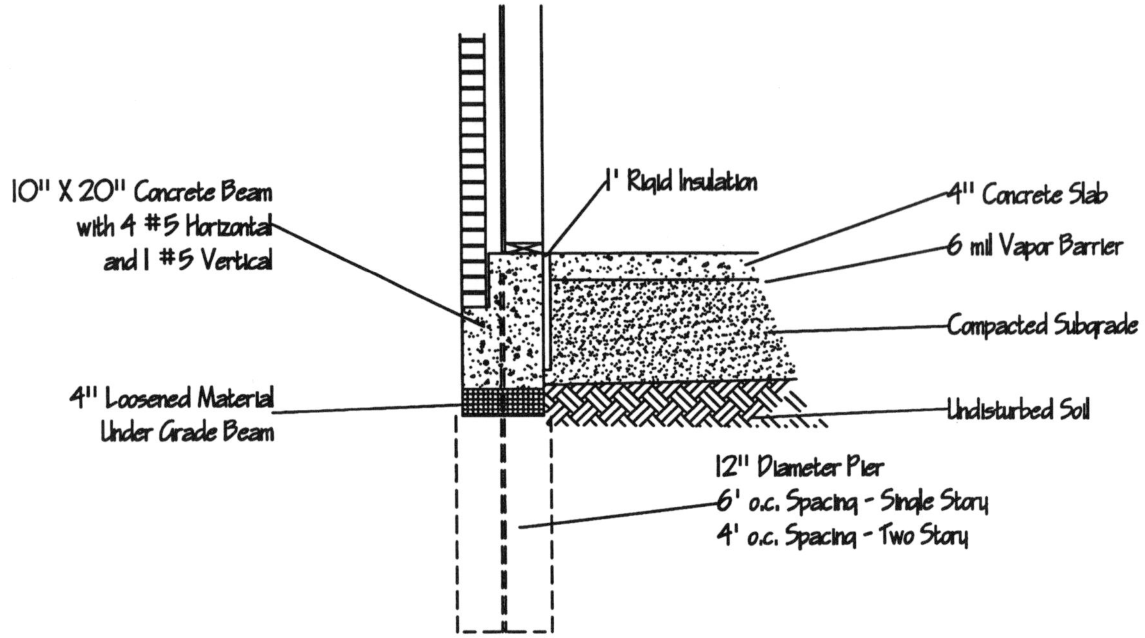

Calculating the Diameter and Placement of the Piers

The diameter of the piers used in residential structures may vary from 12 to 24 inches. The further apart the piers are spaced and the heavier the building, the larger the piers. Piers drilled into a layer of stable dense soil can be smaller than those placed in a lighter soil. For practical reasons, the piers for the entire structure should be the same diameter and drilled so that they bear on the same soil strata. A section of a pier is shown in Figure 12-2.

Some piers are designed with widened bases called *bell piers*. Bell piers require the use of special drilling machines, but the larger base of the bell provides a significantly greater bearing surface, which can reduce the required size of the pier above the bell. A 12-inch-diameter pier with a 24-inch-diameter bottom bell has a bearing capacity similar to that of a 24-inch-diameter pier.

The spacing of the piers also can vary. Most designs space piers every 6 to 12 feet around the exterior foundation walls. Piers also are placed at every corner and at points within the foundation where additional support is needed for interior load-bearing partitions.

The size of the grade beams also affects the spacing of the piers. Larger grade beams can span greater distances but require larger piers. Smaller grade beams span shorter distances but use smaller and more frequently spaced piers. One advantage to using smaller piers is that they can be drilled using attachments, such as an auger, to light tractors and front-end loaders, as shown in Figure 12-3. The tractor shown can be equipped with an auger that can drill 9-, 12-, 16-, and 24-inch-diameter holes to a depth of 42 inches.

FIGURE 12-3. A tractor with auger attachment can drill holes for short piers.

Source: Steiner Turf Equipment, Inc., Dalton, OH.

Reinforcing Piers

Site conditions determine if steel reinforcement is required in the piers. When soils are not compacted or expansive or when the piers' heights are unusually tall compared with their diameter, reinforcing steel may be required to strengthen the pier. Rebar is placed vertically into the piers during the forming process. In large piers, several runs of rebar may be required.

The rebar may be held together with metal hoops that form a cage running the entire height of the pier. The use of cages facilitates the proper placement of the reinforcing steel. Because all the rebar is linked together the rods are less likely to move out of position when the concrete is placed. The rebar also may be extended above the top of the pier and tied to the horizontal rebar in the grade beams as shown in Figure 12-4.

FIGURE 12-4. Rebar may extend above the piers and tie into the rebar that reinforces the grade beam.

Sizing the Grade Beam

Grade beams can vary in thickness from 6 to 24 inches and in depth from 20 and 36 inches. The distance between the piers and the design load of the residence determine the size of the grade beam. Grade beams typically include a minimum of four 5/8-inch #5 rebar. Larger grade beams may incorporate more reinforcing rebar tied together with rods of a smaller diameter called *stirrups*.

The rebar in the grade beams is tied to the rebar that extends up from the top of the piers, securing the two pieces together. To keep the reinforcing steel level while the grade beams are cast, the rebar in the grade beams may rest on chairs or be tied to steel pins or short pieces of rebar that have been driven into the soil at intervals along the length of the beam.

Casting the Pier

The first step in constructing a pier-and-grade-beam foundation is to mark the location of the piers. Set batter boards to provide reference points, but make sure they are far enough from the piers so that they do not disturb the surrounding soil. Use small plastic flags on stiff wire or nails with a ribbon attached to mark the exact location of the center of each pier. Because the drilling machines will disturb the soil on the site, set up reference points outside the immediate drilling area that can be used to measure and confirm the placement of the piers.

Exact placement of the piers is essential. If the pier does not line up properly under the grade beam, soil expansion or contraction can exert enough pressure to push the beam off the pier. It also is essential that the holes for the piers be drilled precisely vertical. Piers that are cast out of plumb or are misshapen can seriously weaken the foundation system.

Because truck- or tractor-mounted drill rigs can be cumbersome to move around a building site it is a good idea for a supervisor to be onsite during the drilling operation to monitor and confirm the correct placement of the piers.

Once the drilling is complete place the required reinforcing steel. Center the rebar cages within the holes and brace them at the top by inserting a piece of wood or rebar through the cage and tying it to stakes driven into the ground. Then tie the rebar in the cage to this support to secure it in place.

If the upper portion of the hole for the pier is too wide or if the top of the pier will be above ground level, construct a form for the top of the pier. Round cardboard tubes in various diameters are available from most lumberyards for this purpose. The tubes generally are left in place after the pier is cast and will, in time, disintegrate. In expansive soils, it is important that the top part of the piers be carefully formed. Bulges at the top of a pier can be lifted by expansive soils.

Usually concrete with a minimum strength of 2,500 psi and no more than a 6-inch slump is recommended for piers. Begin the casting by placing a 4- to 6-inch-deep layer of concrete into the bottom of the pier, taking care not to drop the concrete a distance greater than 4 feet. See Chapter 5 for systems that minimize the drop height.

Casting the Grade Beams

After all the piers are cast, place the concrete in the grade beams. This cast can start as soon as the concrete has been placed in the piers. Some builders erect the forms for the grade beams before the piers are cast, and then cast the piers and grade beams simultaneously. While one casting should take less time than two, there are considerations that may eliminate the time advantage.

To cast both the piers and grade beams together, dig the piers first and then set the forms for the grade beams. One drawback to this method is that it leaves the holes for the piers open longer. In cohesive soils the sides of the holes should stay together, but with granular soils the sides could begin to sluff off. Workers also might unintentionally knock soil into the opening or their weight might cause the sides of the hole to collapse. If this happens it may be necessary to clean out the bottom of the holes before placing the rebar and concrete. Weather conditions, such as rain, also might weaken the soil and make it necessary to cast the piers separate of the grade beams.

If the soil is stable the grade beams may be cast in a trench dug into the earth. Place the steel reinforcing rods in a manner similar to that described for footings in Chapter 6. If the site has unstable soils, the grade beams usually are formed above the surface with site-built plywood forms. Brace the sides of the forms securely to assure that the sides of the beam are plumb. Any deviations from straight lines will mar the appearance once the framing is in place. The forms for a grade beam with a fireplace knock out are shown in Figure 12-5. Bracing for grade-beam forms is shown in Figure 12-6.

If a brick or stone veneer is specified cast a ledge at least 4 inches lower than the top of the grade beam to support the veneer. The ledge also will collect moisture that condenses behind the veneer and channel it to the exterior through weep holes rather than under the sill plate and into the structure. If a masonry veneer is used with this system, the grade beams may be exposed. Use landscaping to hide the exposed beam. To form the ledge, fasten 2×4 blocking in the form to create the ledge, as shown in Figure 12-7. Remove the blocking after curing along with the remainder of the forms.

Because grade beams span from one pier to the next, it is not necessary for the beams to be cast on compacted soils. In fact, in areas where the potential exists for minor amounts of soil expansion, it is

FIGURE 12-5. Forms for grade beams may include blockouts, such as this one for a fireplace.

FIGURE 12-6. Brace the forms for the grade beam before placing the concrete.

preferable to cast the beams on soil that is not compacted. Some localities even require that the area under the beam be excavated and back filled with loose soil before the beams are formed and cast.

Certain soils can expand to fill gaps greater than 6 inches under grade beams. In these situations special corrugated cardboard *fillers* are placed under the grade beams before casting to ensure the foundation's long-term integrity. These fillers are designed to occupy the space under the grade beams while the foundation is constructed. The fillers compress as needed to accommodate soil expansion and eventually disintegrate, leaving an open space under the grade beams.

Before casting the grade beams brush away all loose soil from the top of the cast piers. Apply a chemical bonding agent to the surface of the pier to enhance the connection between the surface of the pier and the grade beams. Then cast each beam in a continuous pour to assure that a cold joint does not form. If necessary, rod layers of concrete with a shovel or tool handle to prevent a cold joint. Place anchor devices in the surface of the grade beam before the concrete begins to set.

After the concrete has cured, remove the forms. This usually can be done the day after casting in warm weather, but during colder seasons sufficient curing could take several days. Once the forms are removed, set the plumbing and mechanical system and construct the floor using a slab or traditional wooden floor system.

Installing Post Foundations

Although more commonly used for garages and barns post foundations can be an attractive alternative for home construction. Post foundations transfer the load through a concrete footing or pad at the base of the hole. As with piers, the greater the diameter at the base of the hole the more the load is spread. The load-bearing capacity of the soil determines the diameter of the hole. A post foundation works best in stable soil with a high load-bearing capacity.

Place concrete in the bottom of the hole, to a depth of at least half the diameter of the hole. After the concrete sets, usually within 24 hours, place a treated wood post on the concrete pad. Fill the remaining void in the hole with dirt lifts. Ram the lifts with a blunt rod or timber or compact them with a mechanical compactor. Repeat the process until the hole is filled and the dirt compacted.

FIGURE 12-7. Use 2x4 blocking to cast a brick ledge in the grade beam.

Exercise care during the backfilling process to avoid moving the post out of alignment.

Once the posts are embedded, attach wooden beams that span from post to post with joist hangers and nails designed to meet code requirements. Then attach floor joists to the wooden beams. A foundation plan for a post-framed building is shown in Figure 12-8. A section of a post-framed building is shown in Figure 12-9.

Driving Pilings

Pilings are vertical members made of steel, concrete, or treated wood that are mechanically driven into the soil. Two basic types of pilings are used in residential construction. *Point-bearing* pilings are driven to a predetermined depth, usually through unstable soil to a layer that can support the load of the residence. *Friction* pilings are driven into the ground until a predetermined bearing capacity is obtained or until the pilings refuse to go any deeper.

The procedure for placing pilings basically is the same as that for laying out piers. If the site will have restricted access, it usually is best to refer back to reference points outside of the disturbed area for the precise placement of each piling.

Driving a piling into the ground is similar to driving a nail into wood. The bearing capacity is determined by the number of blows required to drive the piling a set distance. The contact between the surface of the piling and the soil holds the piling in place just like friction holds a nail in a piece of wood.

FIGURE 12-8. The foundation plan for a post-framed building specifies the size and placement of the pressure-treated timbers.

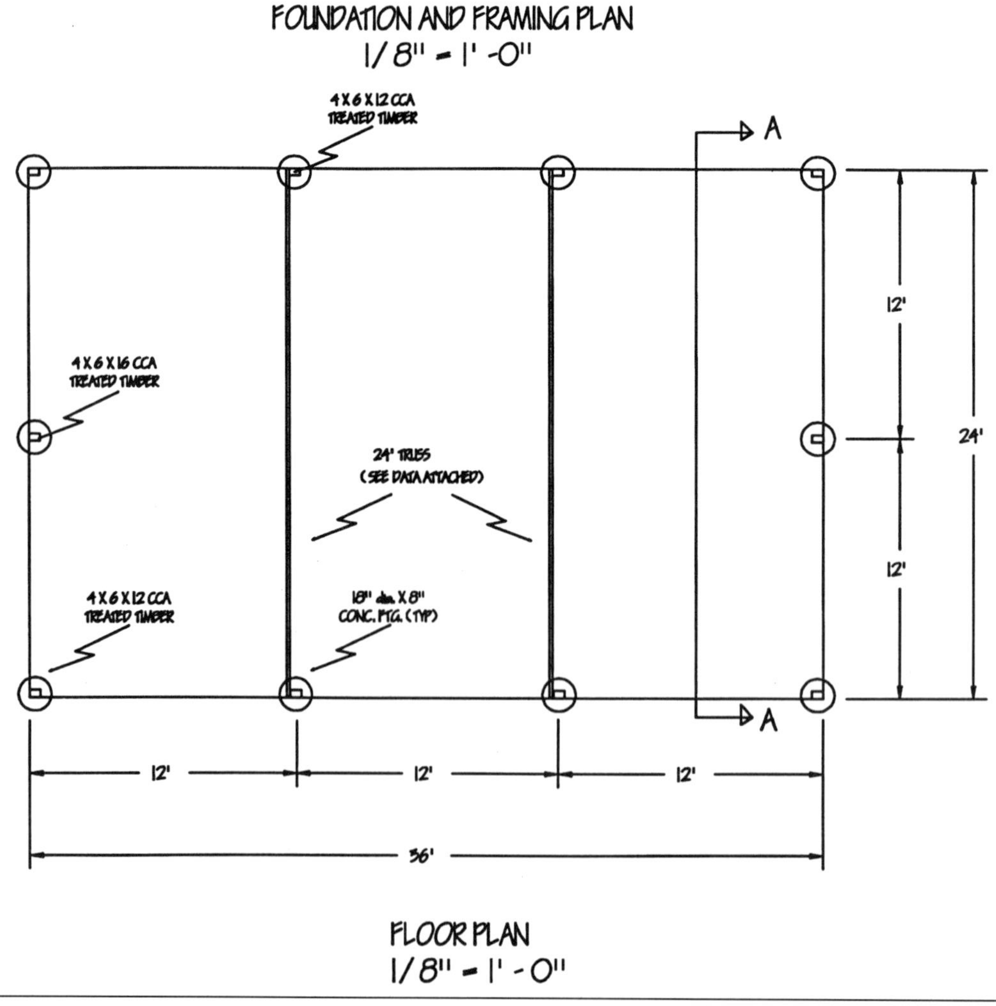

FIGURE 12-9. This section of a post-framed building illustrates the concrete footing on which the post sits, and the splashboard, which is part of the drainage plan.

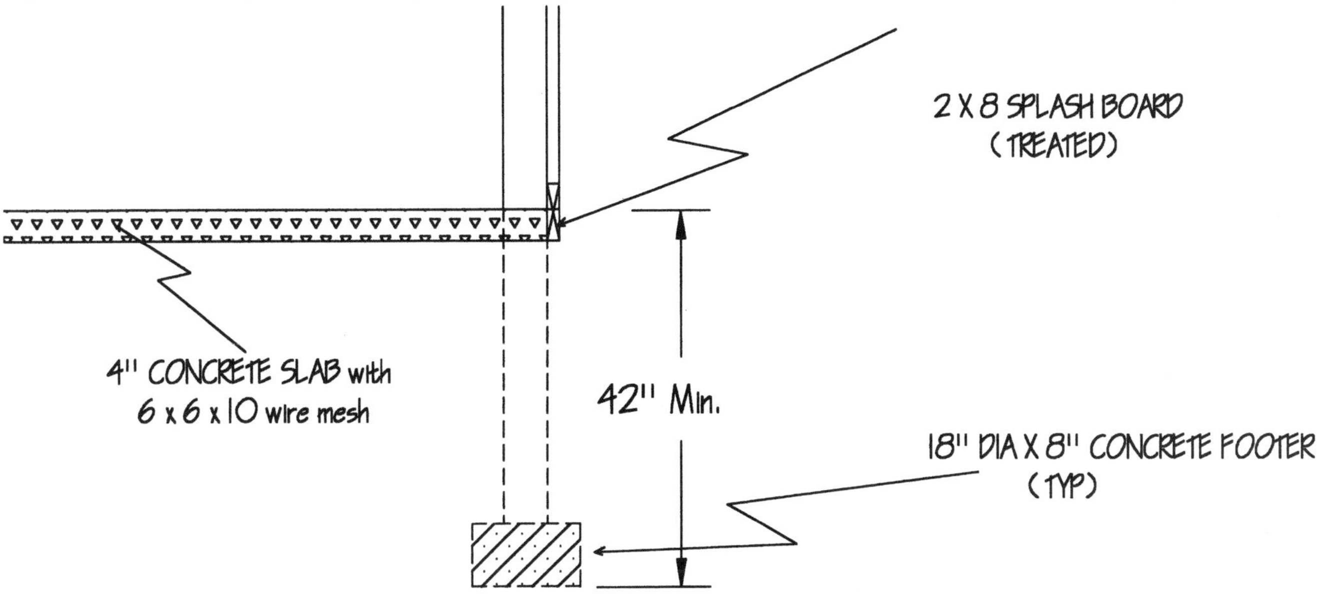

Pilings most often are used along shorelines or in areas where the home must be lifted off the ground. They usually are connected to wood beams in residential construction although, like piers, they may be connected to concrete grade beams. As with piers, the spacing of the pilings is determined by the residence's shape and number of stories, and site conditions such as the type and load-bearing capacity of the soil.

Wood pilings treated with a retention rate of 1 pound of chromated copper arsenate (CCA) per cubic foot are driven mechanically into sandy soil until the friction between the earth and the pilings reaches an adequate force to support the structure or until the pilings reach the desired depth. Usually the pilings are designed so that 8 to 10 feet of timber extend above the ground to support the home above the high-water mark.

Steel pilings for residential foundations usually are placed using one of two methods. In one method, 1 1/2- to 2 1/2-inch-diameter steel rods are pushed hydraulically into the earth until they stop or until the friction between the earth and the rods reaches a predetermined level. Steel pilings pushed deep into the ground should not deteriorate significantly in normal soils, but in corrosive soils they should be galvanized. Similar systems use heavy walled pipe up to 4 1/2 inches in diameter to push the steel pilings into the earth. The rods generally are 10 to 20 feet in length but can be screwed together to create pilings of significant length. A piling being driven into the ground is shown in Figure 12-10.

The second method uses 1 1/2- to 2 1/2-inch square rods with annealed steel plates at one end. The plates act as an auger to screw the pilings into the ground. When the force required to turn a piling into the earth reaches a predetermined level, the piling is deemed to be deep enough. Figure 12-11 shows the anchor tool that permits a hydraulic jack to transfer the load to the bracket and foundation anchor.

Both methods use a driver that can be attached easily to the boom of a backhoe. Because franchised operators usually install these systems, they are not necessarily available in all areas of the country.

In some cases, steel pilings are driven into the bottom of a hole drilled for a pier. After the steel piling is driven to the desired depth or resistance, a concrete pier is cast around the steel piling. This creates a hybrid system that is stronger than either system used alone. This combination foundation works well with expansive soils. The steel piling can be driven to bear on stable soils deep below the surface and the pier part of the foundation provides a stable base that supports the grade beam.

After the steel pilings are in place, grade beams are cast to span the distance between the pilings. The

FIGURE 12-10. Pilings are driven into the ground until they reach a specific depth or the friction between the earth and the pilings reaches a predetermined level.

Source: Hickson Corporation, Smyrna, GA.

rebar in the grade beams is attached to brackets secured to the pilings.

Chapter Checklist

Consider the following questions when working with pier, post, and piling systems.

- Is each pier, post, or piling accurately placed and the center marked?
- Are reference points marked in undisturbed areas to help position the pier, post, or piling?
- Is the equipment set up to ensure that piers will be drilled vertically and that the holes will have straight sides?
- Are cardboard tubes available to form the top of the above-grade portion of the piers?
- Are precautions in place to keep workers and equipment from disturbing previously drilled piers while drilling or casting the remaining piers?
- Do the grade beams conform to the plan? Are they squared and of the proper thickness and depth?
- Has reinforcing steel been placed according to the engineering requirements?
- Is the brick ledge formed per the plan?
- Is the concrete at the correct slump without added water?
- Are the pilings driven to the specified depth or to the predetermined resistance?
- Are the posts vertical and centered on the concrete pads that serve as a footing?
- Has all lumber that will be in contact with the ground been treated with the proper preservative at the correct retention rate?

FIGURE 12-11. An anchor tool permits the hydraulic jack to transfer the load to the bracket and foundation anchor.

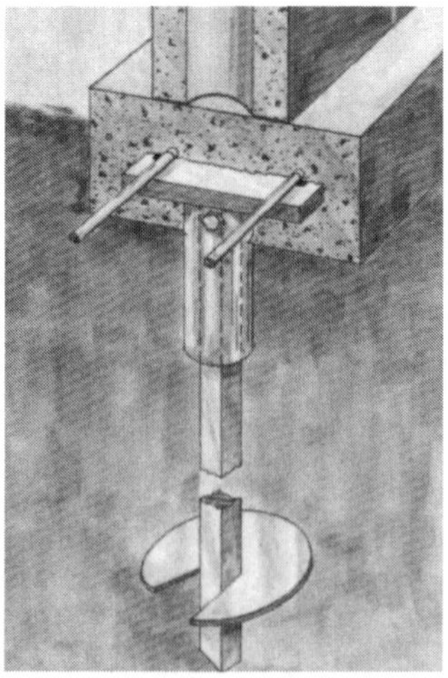

Source: A.B. Chance Co., Centralia, MO.

13

Building Preserved Wood Foundations

Preserved wood foundations have several advantages. Construction is not limited to the warm months of the year because wood is not affected by extreme hot or cold weather. There are no delays waiting for cement to cure or forms to be removed. Subcontractors that specialize in concrete and other foundation types are not needed, potentially saving time and reducing costs.

Installation of the electrical and plumbing systems also is easier. The studs used to erect the walls can be drilled easily and fixtures secured in place. The foundation walls are easily insulated and finished, bypassing the need to attach furring strips.

Perhaps the greatest advantage is that owners of homes with a preserved wood foundation report that the space has a comfortable feel. Because a drainage system is built into the foundation, basements do not feel damp and moist, but dry.

Often used to construct forms and temporary bracing, wood became a material of choice for foundations with the advent of water-borne preservation chemicals that provide effective protection from insects and decay-causing fungi. Unlike earlier preservation techniques that used coal tars and creosote, today's treated wood products can be sawn and nailed as easily as untreated lumber. Contact with the skin is not harmful and the treated lumber does not emit noxious gases as did wood treated with earlier preservation chemicals.

Nonetheless, some home buyers are leery of this type of foundation. It's difficult to understand that wood, buried in the ground as part of a foundation system, does not decay or become infested with termites. Most manufacturers of pressure-treated foundation wood warranty the wood to last a lifetime. Tests conducted on sample treated stakes that have been in the ground approximately 50 years show that the preservation treatment has maintained the integrity of the wood.

While today's preserved wood is safe, care should be taken when working with all treated lumber. Workers should wear respiratory protection and goggles when sawing and driving nails. Do not burn scrap lumber pieces because the chemicals remain in the ashes and can enter the groundwater. The most environmentally friendly way to dispose of scrap is to bury the pieces because the chemicals will not leach out of the lumber.

Understanding the Pressure Treatment Process

All wood that will be a permanent part of a foundation system should be pressure treated, including 1x boards, dimensional 2x lumber, and plywood. Each piece also must meet grade requirements and dimensions established by local building codes.

All lumber is marked with a grade stamp that indicates the species, production mill, and information about the strength of the piece of lumber, as shown in the grade stamp in Figure 13-1. In addition, each piece of lumber and plywood sheet must be stamped to show the concentration of treatment. The stamp indicates the agency certifying the treatment, the location of the treating facility, and the year of the treatment. Some producers also tag each piece of lumber and plywood with a label that contains similar information.

FIGURE 13-1. The grade stamp on pressure-treated lumber indicates the information required by building codes, including the species, production mill, and concentration of treatment.

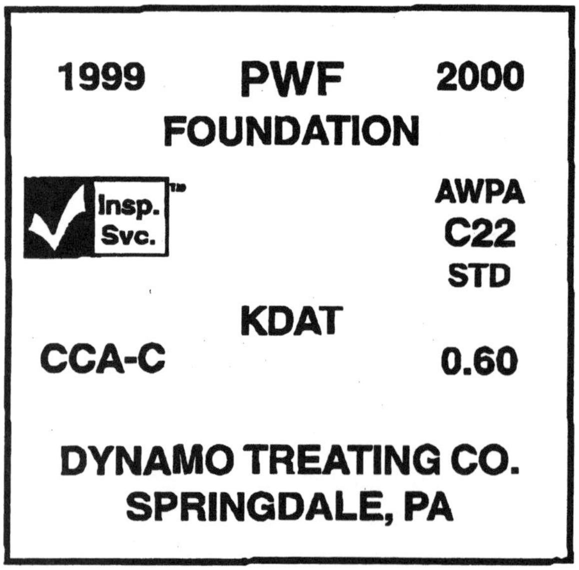

Source: Hickson Corporation, Smyrna, GA.

Lumber intended for use in foundations also must be stamped foundation grade (FDN) or indicate that the amount of the chemical concentration is at least 0.60 pounds per cubic foot (pcf), which is 50 percent more than "ground-contact" requirements. To minimize shrinkage, FDN-grade lumber also may be kiln dried after treatment.

The amount of the concentration, or the *retention rate*, of the chemicals in a given batch of wood is stated in terms of the pounds per cubic feet of the water-borne salts retained in the wood. The American Wood Preservers Bureau (AWPB) and the American Wood Preservers Association (AWPA) have established standards for the chemical concentrations with which the wood should be treated for ground-contact use. Ground-contact-rated lumber must have at least a 0.40 pcf retention rate. Treating facilities are highly regulated both by internal associations such as AWPB and AWPA and also by the Environmental Protection Agency (EPA). Strict regulations control the use of the preserving chemicals and the design of the treating facilities

During the treatment process chemicals are forced into the cells of the wood. Finished lumber is loaded onto carts and placed in large pressure vaults. The vault is sealed and the air inside the chamber is removed to create a partial vacuum. The chamber then is filled with the preservatives. Chromated copper arsenate (CCA), the most widely used treatment, is a solution of the water-borne salts of chromium trioxide, copper oxide, and arsenic pentoxide. The CCA solution is pumped into the cylinder until the required pressure is obtained in the treating chamber. This pressure is maintained for a calculated time period determined by the species, quantity of lumber being treated, and the desired concentration of preservatives. After the lumber is treated the chamber is emptied.

A vacuum then may used to remove the remaining CCA solution in the chamber that was not absorbed by the lumber. To check the concentration of the treatment, the amount of chemicals remaining in the holding tanks is measured and compared to the quantity in the tanks before treatment. The difference is a fair indication of the extent of the treatment. Test borings of the treated wood also can be made to measure the concentration.

Other chemicals besides CCA are used to pressure treat lumber and plywood used in residential construction. These include ammoniacal copper arsenite, chromated zinc chloride, and fluorochrome arsenate phenol.

Providing Adequate Drainage

Adequate drainage is necessary to prevent water-related problems in preserved wood foundations. To support drainage most builders use a gravel footing. The footing should be designed to meet onsite soil conditions and be approved by local building officials. See Chapter 6 for information on the installation of gravel footings. Most footings are twice as wide and three-quarters as deep as the width of the footing plates, which are the bottom wood members of the foundation walls.

Building Foundation Walls

Begin building the foundation walls by placing the *footing plate*. This piece of treated dimensional lumber should be at least 2 inches wider than the studs that will be used to build the walls. Some builders secure the footing plate to the gravel bed by driving 12- to 18-inch-long pieces of rebar into the footing along the sides of the plate. This practice keeps the footing plate from shifting while the wall is being built. Mark the perimeter of the residence on the footing plate using a chalk line.

The size and spacing of the studs and the exterior sheathing for the foundation walls usually are determined by the number of stories in the structure and the amount of backfill from the footing to the finished grade. Section 404.3.2 of the Council of American Building Officials (CABO) (95) code requires that the 2×6 studs used in foundation walls with a backfill of 48 inches or less have a minimum *force of bending* (f_b) of 1,250 when placed 16 inches on center and 875 f_b when placed 12 inches on center. These values can be used to determine the grade and species of lumber needed to meet the code's engineering requirements. According to the table, #2 southern yellow pine meets these requirements. If the foundation will be back filled to a depth greater than 4 feet, the wall must be engineered to meet the specific site conditions. More than likely this will require that the studs be increased to 2×8s. Figure 13-2 shows the typical details for wood foundation basement walls.

Attach the studs using stainless steel nails to the top and bottom plates, which should be the same size as the studs. Apply sheathing of FDN-grade pressure-treated plywood to the exterior of the foundation walls and affix with stainless steel nails or screws and an adhesive approved for below-grade use in foundations. Then erect and nail the wall panels to the footing plate, plumb them, and provide adequate bracing.

Carefully caulk all joints between the sheets of plywood to prevent moisture from penetrating the system. The plywood may be aligned vertically with the studs or placed horizontally, perpendicular to the studs. Figure 13-3 provides the grade, minimum thickness, and identification number of the plywood based on its alignment with the studs. If the plywood is placed perpendicularly to the studs, place *blocking* made of treated dimensional lumber between the studs to support all four edges of the plywood. Construct headers above all window and door openings from treated wood of an adequate thickness to support all imposed loads. Lumber sizes for the headers are provided in Figure 13-4.

Leave the bracing supporting the walls in place until the basement floor and first floor are constructed and waterproofing applications and back filling are complete. The floors help absorb the lateral stress exerted on the foundation walls by the backfill. If the backfill exceeds 48 inches, install additional blocking between the joists that are parallel to the foundation walls to ensure adequate support.

The rim joists on walls that are parallel to the floor joists also may require bracing so they don't twist and move the top of the foundation walls. For every 12 feet of foundation wall, provide one set of bracing from the rim joist inward to at least the fourth joist. In high-wind areas, mechanically attach the members of the flooring system to the studs of the foundation system with straps.

Prior to back filling, cover the exterior side of the foundation wall with a waterproofing compound or a sheath of 6-mil polyethylene film. If film is used, cover it at and just below the level of the finished grade with a piece of treated lumber or plywood to keep the film from being damaged.

Putting in the Floor

The basement floor of an all-weather wood foundation can be either a 4-inch concrete slab or a wood floor constructed of approved treated joists and plywood sheathing. Build either floor over a 4-inch-deep compacted gravel subgrade. Treat the soil to prevent the infestation of insects such as termites and place a vapor barrier of 6-mil polyethylene film over the gravel bed.

Place and finish the concrete floor when the top of the gravel fill is even with the top of the footing plate on the foundation wall. To make it easier to screed a concrete slab some builders nail a 1× screed board to the wall studs to mark the top of the basement slab. This board also prevents concrete from filling the space

FIGURE 13-2. This detailed drawing illustrates the required elements for a wood foundation wall.

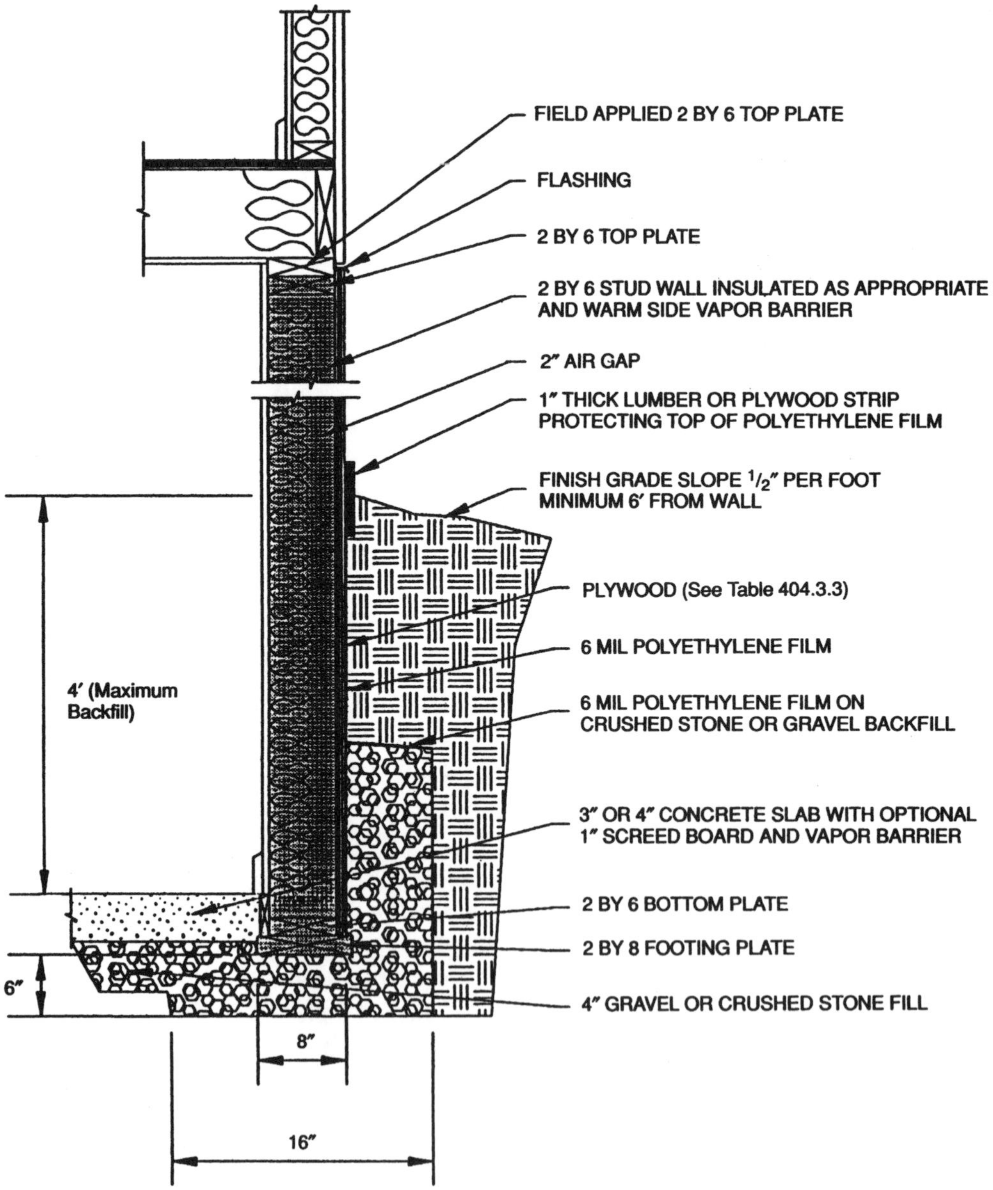

For SI: 1 inch = 25.4 mm, 1 foot = 304.8 mm.

Source: Council of American Building Officials (95) Figure 403.1b.

FIGURE 13-3. Plywood grade and thickness is based on standard calculations.

HEIGHT OF FILL (inches)	STUD SPACING (inches)	FACE GRAIN ACROSS STUDS			FACE GRAIN PARALLEL TO STUDS		
		Grade[1]	Minimum Thickness	Identification Index	Grade[1]	Minimum Thickness[2,3]	Identification Index
24	12	B	$^{15}/_{32}$	32/16	A	$^{15}/_{32}$	32/16
					B	$^{15}/_{32}$[3]	32/16
	16	B	$^{15}/_{32}$	32/16	A	$^{15}/_{32}$[3]	32/16
					B	$^{19}/_{32}$[3] (4, 5 ply)	40/20
36	12	B	$^{15}/_{32}$	32/16	A	$^{15}/_{32}$	32/16
					B	$^{15}/_{32}$[3] (4, 5 ply)	32/16
					B	$^{19}/_{32}$ (4, 5 ply)	40/20
	16	B	$^{15}/_{32}$[3]	32/16	A	$^{19}/_{32}$	40/20
					B	$^{23}/_{32}$	48/24
48	12	B	$^{15}/_{32}$	32/16	A	$^{15}/_{32}$[3]	32/16
					B	$^{19}/_{32}$[3] (4, 5 ply)	40/20
	16	B	$^{19}/_{32}$	40/20	A	$^{19}/_{32}$[3]	40/20
					A	$^{23}/_{32}$	48/24

For SI: 1 inch = 25.4 mm, 1 pound per cubic foot = 0.1572 kN/m^3.

1 Plywood shall be of the following minimum grades in accordance with DOC PS 1 or DOC PS 2:

(i) DOC PS 1 Plywood grades marked:
  a. Structural I C-D (Exposure 1)
  b. C-D (Exposure 1)

(ii) DOC PS 2 Plywood grades marked:
  a. Structural I Sheathing (Exposure 1)
  b. Sheathing (Exposure 1)

(iii) Where a major portion of the wall is exposed above ground and a better appearance is desired, the following plywood grades marked Exterior are suitable:
  a. Structural I A-C, Structural I B-C or Structural I C-C (Plugged) in accordance with DOC PS 1
  b. A-C Group 1, B-C Group 1, C-C (Plugged) Group 1 or MDO Group 1 in accordance with DOC PS 1
  c. Single Floor in accordance with DOC PS 2

2 Minimum thickness $^{15}/_{32}$ inch, except crawl space sheathing may be $^{3}/_{8}$ inch for face grain across studs 16 inches on center and maximum 2-foot depth of unequal fill.

3 For this fill height, thickness and grade combination, panels which are continuous over less stud spacings require blocking 16 inches above the bottom plate. Offset adjacent blocks and fasten through corrosion-resistant nails at each end.

Source: Council of American Building Officials (95) Table 404.3.3.

FIGURE 13-4. Lumber sizes for headers are based on standard calculations.

SIZE OF HEADER[1,2]	HEADERS IN BEARING WALLS[2]			HEADERS IN WALLS NOT SUPPORTING FLOORS OR ROOFS
	Supporting Roof Only	One Story Above	Two Stories Above	
2 × 4	4	—	—	—
2 × 6	6	4	—	—
2 × 8	8	6	—	10
2 × 10	10	8	6	12
2 × 12	12	10	8	16

For SI: 1 inch = 25.4 mm, 1 inch = 304.8 mm.

1 Nominal 4-inch-thick single headers may be substituted for double members.

2 Spans are based on No. 2 Grade Lumber with 10-foot tributary floor and roof loads.

Source: Council of American Building Officials (95) Table 602.6.

between the studs at the base of the foundation, but does not weaken the system. Chapter 10 includes information on placing and finishing concrete slabs.

The procedure for installing a treated wood basement flooring system is similar. Place the compacted gravel bed, treat the soil, and install the vapor barrier. The pressure-treated floor joists can bear directly on the footing plate. Because the joists will rest on gravel there is no distance to span so 2×4s can be used for the joists.

After the joists are placed, attach the plywood sheets. The thickness of the plywood is determined by the spacing of the joists. Plywood is stamped to indicate acceptable spans. For example, plywood stamped 32/16 is rated to span gaps of 16 inches in flooring systems. The number 32 means the plywood is rated to span 32-inch spaces when used on roofs.

As with the construction of the foundation walls all lumber and plywood used in the construction of the flooring system must be treated for ground-contact use. All nails should be stainless steel and all adhesives approved for use in these flooring systems.

Installing Electrical and Plumbing Systems

It is easy to install electrical and plumbing systems, as well as insulation, on the inside of a preserved wood foundation. To avoid weakening the studs, drill the holes needed to run wires and plumbing lines in the center of the studs. Holes drilled in the center of studs do not diminish their structural integrity, but notches cut in the edges of the studs do. Do not use notches to place piping.

Constructing Crawl Spaces

Crawl spaces also can be constructed using preserved wood. As with a basement foundation, construct the crawl space on a level, well-compacted gravel footing, the top of which is below the frost line. Erect the walls on a footing plate using the same methods and criteria used to build a full basement wall. Within the crawl space, the top of the foundation wall must be at least 8 inches above the finished exterior grade. The bottom of the floor joist must be a minimum of 18 inches above the finished grade.

The major difference between basement walls and crawl-space walls is the placement of the exterior sheathing. In a crawl space, pressure-treated plywood exterior sheathing is required only from the top of the foundation wall to a depth that is 10 inches below grade. To counter deflecting forces exerted by the soil, support the bottom edge of the below-grade sheathing by placing blocking between the studs. Figure 13-5 shows a typical detail for wood crawl-space walls.

After the foundation walls are in place and braced to prevent movement, backfill both the interior and exterior sides of the foundation so the earth fills the cavities between the studs and secures the walls in place. Then remove the bracing

If posts are needed to support interior beams, center and support them on their own footing plates. The footing plates, made with 2× treated lumber, must extend at least 2 inches beyond the perimeter of the post in each direction. For example, a 4×6-inch post would center on at least an 8×10-inch footing plate. Each footing plate must rest on a gravel bed at least 30 inches in diameter and 8 inches deep.

It is best to place the gravel for these posts when placing the gravel for the footing. Gravel beds for interior footings do not need to be placed below the frost line but they must bear on undisturbed soil. The spacing of the posts is determined by the size of the beam needed to support the floor joist. These spans are determined using traditional span charts provided in all building codes. The table in Figure 13-6 can be used to determine the allowable span for girders.

Installing a Vapor Barrier and Insulation

Except where the soil is exceptionally dry and when not required by local codes, a vapor barrier must be installed so that the soil inside the crawl space is covered. A polyethylene film usually is sufficient to retard the passage of moisture or soil gases into the structure.

Batts placed between the floor joists usually are adequate to insulate the underside of the floor above the crawl space. Local building codes give minimum R-values for this insulation. An alternative is to place insulation in the stud cavities of the preserved wood walls. Extend this insulation down to the frost line and use a product that is very resistant to moisture absorption.

FIGURE 13-5. Typical details for wood walls in crawl spaces are shown in this drawing.

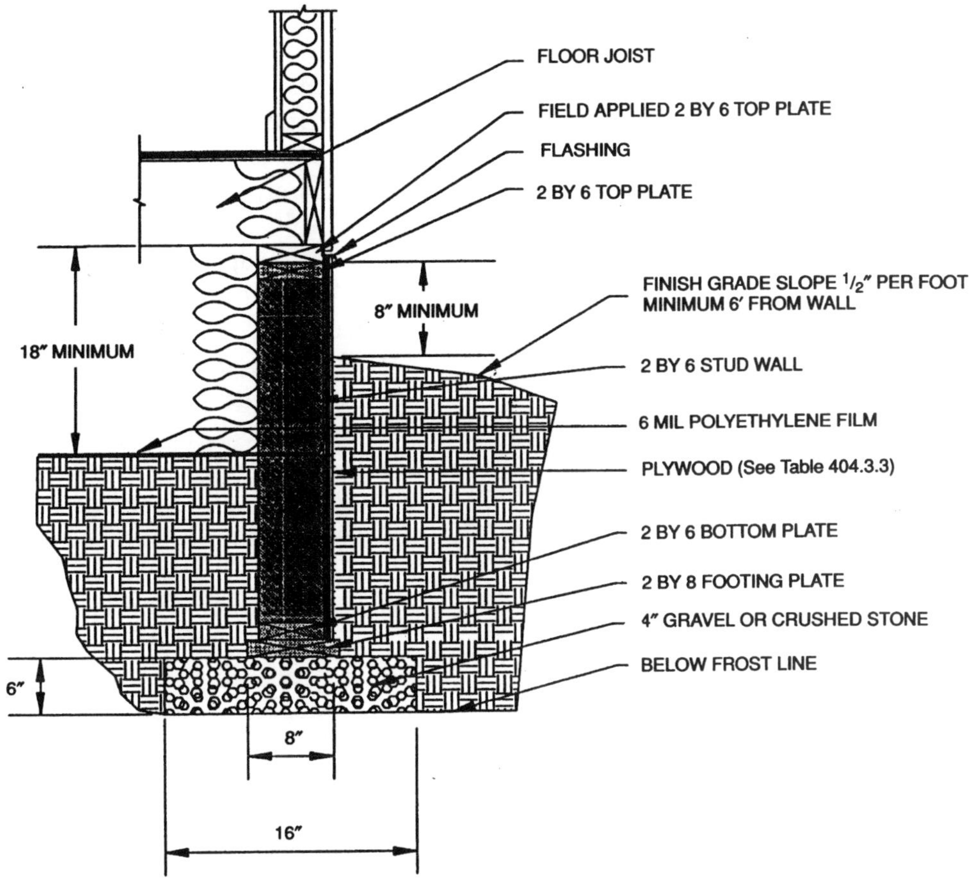

For SI: 1 inch = 25.4 mm, 1 foot = 304.8 mm.

Source: Council of American Building Officials (95) Figure 403.1c.

FIGURE 13-6. The allowable span for girders is calculated based on standard sizes.

SIZE OF WOOD GIRDER[2]		FLOOR LIVE LOAD (psf)	SPACING BETWEEN GIRDERS OR BETWEEN GIRDERS AND LOAD-BEARING WALLS[3]				
			4 feet	6 feet	8 feet	10 feet	16 feet
4 × 4	—	30	5′-6″	4′-6″	3′-6″	3′-0″	2′-6″
		40	5′-0″	4′-0″	3′-6″	3′-0″	2′-6″
4 × 6	—	30	8′-0″	6′-6″	5′-6″	5′-0″	4′-6″
		40	7′-6″	6′-0″	5′-6″	4′-6″	4′-0″
4 × 8	6 × 6	30	11′-0″	9′-0″	8′-0″	7′-0″	5′-6″
		40	10′-0″	8′-6″	7′-6″	6′-6″	5′-0″
4 × 10	6 × 8	30	14′-0″	11′-6″	10′-0″	8′-6″	6′-0″
		40	13′-0″	10′-6″	9′-6″	8′-6″	5′-6″
4 × 12	6 × 10	30	16′-6″	14′-0″	12′-0″	11′-0″	9′-0″
		40	16′-0″	12′-6″	11′-0″	10′-0″	8′-0″

For SI: 1 inch = 25.4 mm, 1 foot = 304.8 mm, 1 psf = 0.0479 kN/m^2.

1 Allowable spans may be interpolated between tributary loads shown in table. Spans and girder sizes may be computed independently of the above table in accordance with accepted engineering practice.

2 Spans are based on No. 2 lumber.

3 The spacing is the tributary load to the girder. It is found by adding the unsupported spans of the floor structure on each side which are supported by the girder and dividing by 2.

Source: Council of American Building Officials (95) Table 502.3.3a.

Providing Adequate Ventilation

To prevent moisture problems, enclosed crawl-space foundations must have adequate ventilation. One square foot of ventilation space must be provided for each 1,500 square feet of area under the residence. The vents must be constructed of a corrosion-resistant material, such as aluminum. To keep out rodents the vents should be made with holes no larger than 1/2 inch. Some vents are designed to be open permanently; others have louvers that help protect the interior space from blowing rain. The vent's free area, a figure that compensates for loss of air movement by the frame and screen, should be indicated on each vent.

The placement of the vents can be determined by the building design provided sufficient vents are included to accommodate the square footage of the crawl space. Place vents on at least three walls to ensure good air circulation within the crawl space. One wall may be free of vents. Eliminating the vents along the north side of the house, for example, can be beneficial in colder regions of the country.

Chapter Checklist

When building a preserved wood foundation, answer the following questions.

- Is the footing below the frost line?
- Is the footing composed of clean, well-compacted gravel with low levels of dust?
- Is the footing plate at least 2 inches wider than the studs that will be used to build the foundation walls?
- Is each piece of lumber and plywood used in the foundation wall or basement floor stamped to certify that it is treated for foundation use?
- Does all ground-contact lumber meet the minimum CCA requirements?
- Can the material grades be verified before delivery is accepted?
- Will only stainless steel nails and fasteners be used?
- Will all plywood be nailed and glued to the studs? Will all edges of the plywood be secured?
- Will additional blocking between the studs be required?
- Will the foundation wall be tall enough to ensure that any untreated lumber, such as rim or floor joists, are at least 8 inches above the finished exterior grade, and for crawl spaces at least 18 inches above the interior grade?
- Has the foundation been waterproofed and is a subsurface drainage system in place?
- For basement foundations, will both the basement floor and the first floor be in place before any back filling is done? Will bracing be retained until after the backfill is in place?
- Is the finished grade designed to provide adequate surface drainage away from the house?
- Is the soil under the crawl space covered with a vapor barrier of polyethylene film?
- Will the crawl space be ventilated at a ratio of 1 square foot of ventilation space to every 1,500 square feet of area under the residence?

14

Creating Frost-Protected Foundations

Building codes in the Unites States generally require that all footing and foundation systems be built to a depth that is below the frost line. The purpose behind the code requirement is to prevent problems that can occur when moisture in the soil freezes, expands, and causes the footing-and-foundation system to heave.

Since the 1950s Scandinavian scientists and engineers have been developing and perfecting a stable foundation system that does not need to extend below the frost line. The base of the foundation must be only 12 inches below the finished grade. The Scandinavian system is based on the premise that heat, escaping from the interior of the home, can keep the footing and underlying soil from freezing. This system also can be adapted for use in nonheated structures by capturing geothermal energy as a source of heat. Over a million homes and light commercial buildings in Scandinavia have been built using these frost-protected foundations with no resulting foundation failure.

Recent revisions to the building codes permit frost-protected foundations in the United States. This system can be used in two situations in regions with a deep frost line.

1. When the added expense of building a basement is not justified.
2. When one side of the basement is below the frost line, but the bottom of the footing on the walkout portion of the basement is above the frost line.

In the second situation, the alternative to the frost-protected foundation is a stepped foundation. Many municipalities enforce the Model Energy Code (MEC), which includes requirements for insulated foundation walls. The R-value of the insulation required for a frost-protected foundation might not be much more than that required by the MEC.

More detailed information on the construction of frost-protected shallow foundations is available from the National Association of Home Builders National Research Center, Inc.

Placing the Insulation

Each of the foundation types described in this book may become a frost-protected foundation with insulation. The insulation does the most good when it is placed on the exterior side of the foundation where it can keep the heat from escaping from under the residence. Insulation placed on the above-grade portion of exterior walls must be protected from damage, usually with a covering of cementitious material such as stucco.

Insulation placed horizontally at least 12 inches below grade also helps retain the heat in the dwelling. This insulation is required in regions that have an *air-freezing index* greater than 2,500, indicated on the map in Figure 14-1. The exact thickness and the precise distance that the insulation must extend are based on climatic conditions at the site.

Corners are protected by extending the insulation a greater distance away from the perimeter of the structure. The design requirements are based on the insulation's use in a heated structure that maintains a minimum temperature of 64° F. This minimum temperature can be reduced for short periods, but the interior temperature should not be allowed to fall below 41° F. Figure 14-2 is a detailed drawing of a frost-protected foundation. Figure 14-3 provides minimum frost-protected insulation requirements.

Installing a Drainage Layer

Although the frost-protected system is designed for use in colder regions of the country, it must be built on unfrozen ground. Drainage and insulation must be provided under the entire home, not just the footing. A frost-protected foundation is placed on a compacted gravel subgrade that is at least 4 inches deep under the entire structure. This layer of gravel continues under the footing and extends outward to the outside perimeter of any horizontal insulation that might be required. The gravel must be cleaned so that it is free of any dust or debris that might hinder its drainage action. Adequate drainage is an important part of this system. Because moisture is a high conductor of heat, it is best to eliminate its presence from beneath the structure.

Next an 8-mil vapor barrier is placed on top of the gravel to further prevent moisture from migrating into the residence. Place a layer of insulative sheathing of polystyrene, mineral wools, or fiberglass over the vapor barrier. The R-value of the insulation is based of the air-freezing index shown in Figure 14-3. On top of the insulative sheathing place an additional 4-inch layer of washed crushed stone or gravel. Then place any standard foundation and flooring system.

An additional layer of horizontal insulation must extend outward from the base of the foundation in regions that have an air-freezing index greater than 2,500. This layer must be placed on a minimum of 4 inches of compacted gravel. The R-value of this layer is provided in Figure 14-3.

Constructing Walkout Basements

On a sloping site a frost-protected foundation is an economical alternative to stepping the footing below the frost line. For example, a traditional foundation system may be used on the sides of the foundation that are below the frost line and the frost-protected system may be used on the remaining sides. Follow the same gravel base and insulation specifications required for frost-protected systems that extend under the entire structure.

Protecting Unheated Structures

Most of the model codes provide design requirements for residences that are heated. The requirements for foundations of unheated structures, such as garages, are not covered. A traditional foundation or a system that uses insulation to trap the earth's geothermal heat must be designed for under these structures.

Chapter Checklist

When considering a frost-protected foundation, ask the following questions.

- Is a frost-protected system appropriate?
- What is the depth of the frost line?
- What are the MEC requirements for the foundation wall?

FIGURE 14-1. The air-freezing index, illustrated on this map, is based on the cumulative number of days in which the temperature falls below 32°F.

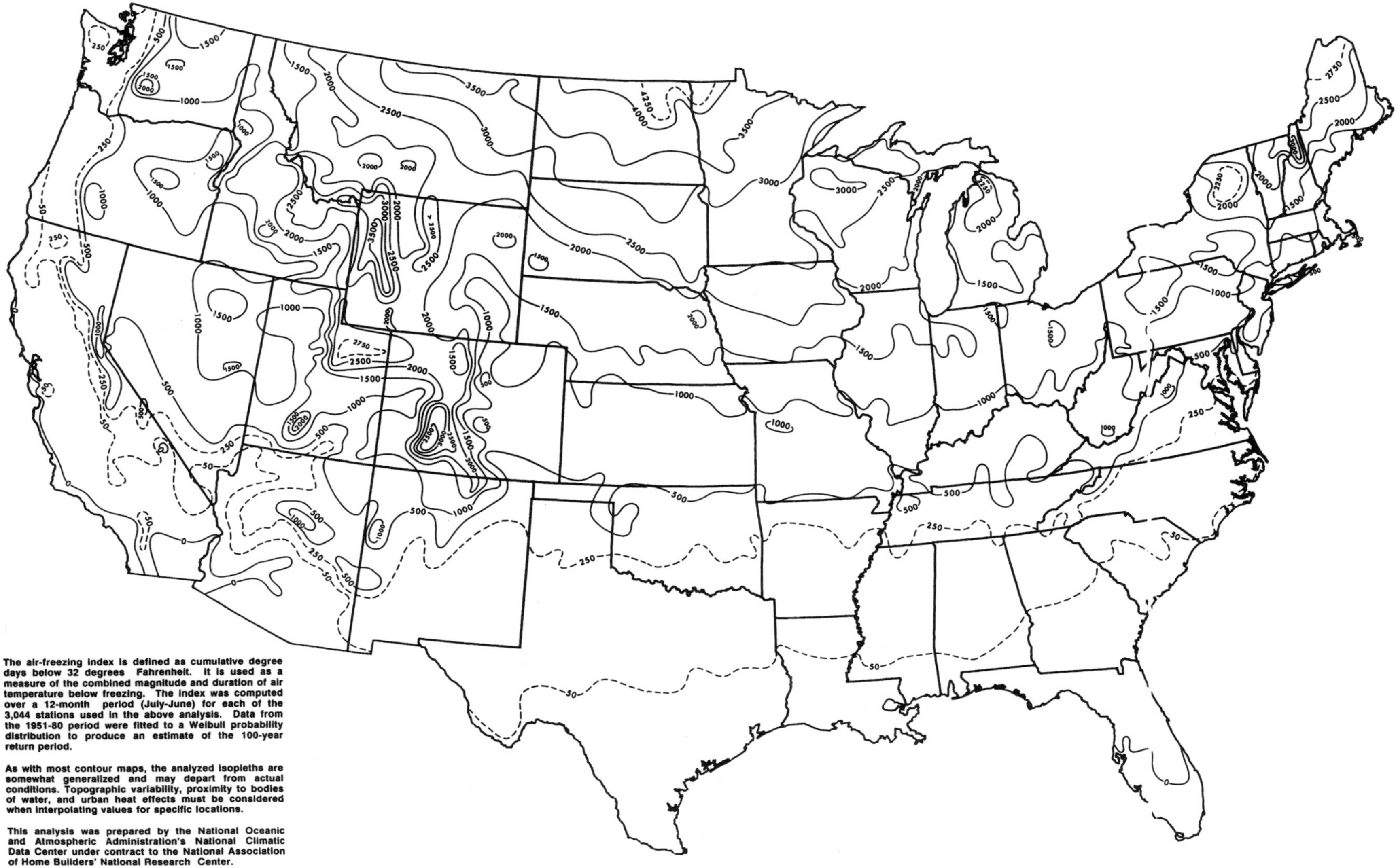

The air-freezing index is defined as cumulative degree days below 32 degrees Fahrenheit. It is used as a measure of the combined magnitude and duration of air temperature below freezing. The index was computed over a 12-month period (July-June) for each of the 3,044 stations used in the above analysis. Data from the 1951-80 period were fitted to a Weibull probability distribution to produce an estimate of the 100-year return period.

As with most contour maps, the analyzed isopleths are somewhat generalized and may depart from actual conditions. Topographic variability, proximity to bodies of water, and urban heat effects must be considered when interpolating values for specific locations.

This analysis was prepared by the National Oceanic and Atmospheric Administration's National Climatic Data Center under contract to the National Association of Home Builders' National Research Center.

Source: Council of American Building Officials (95) Figure 403.3b.

FIGURE 14-2. These drawings illustrate the insulation requirements for a frost-protected foundation.

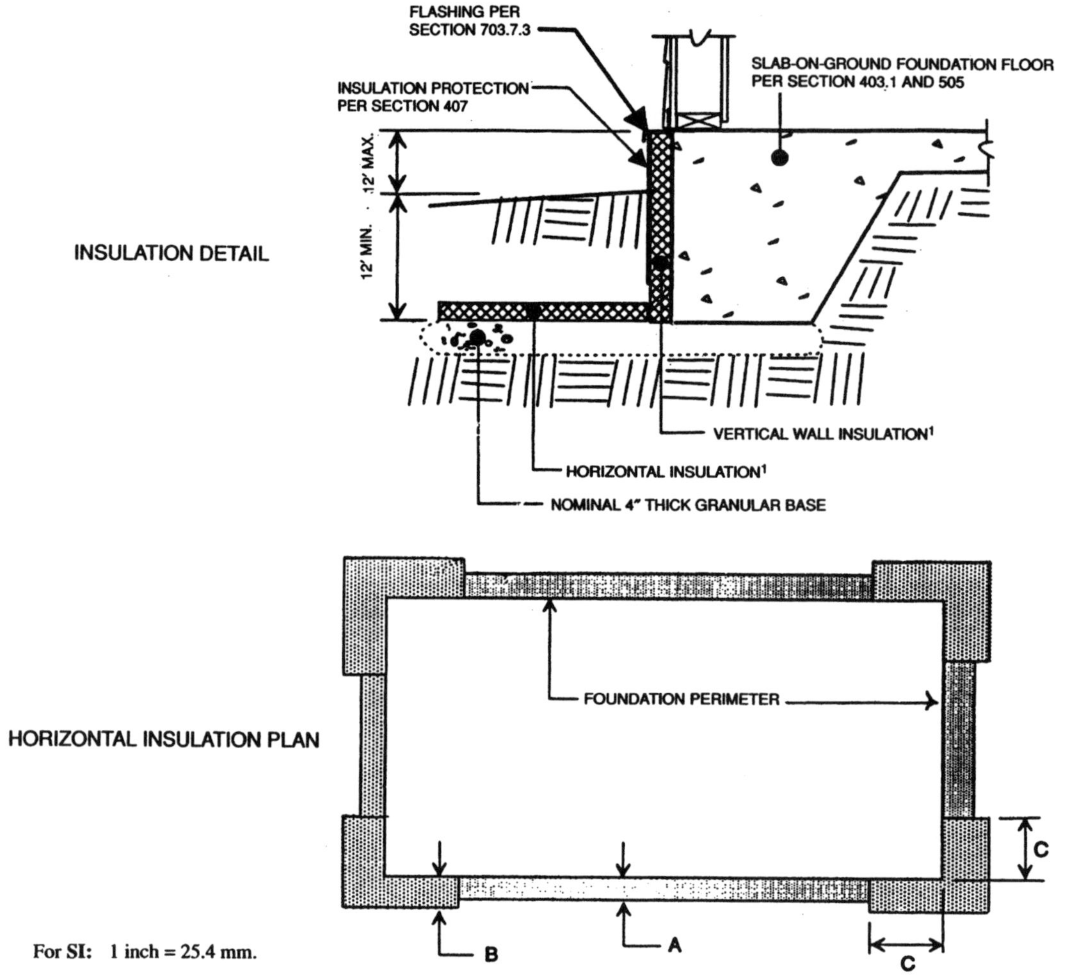

For SI: 1 inch = 25.4 mm.

[1] See Table 403.3 for required dimensions and R-values for vertical and horizontal insulation.

Source: Council of American Building Officials (95) Figure 403.3a.

- Would the cost of the foundation material saved be more than the cost of the added insulation?
- Is the soil at the site frozen?
- What is the air-freezing index for the proposed site?
- Is there an adequate layer of gravel under the structure?
- What are the horizontal insulation requirements?
- What are the insulation requirements for the site?

FIGURE 14-3. The minimum frost-protected insulation requirements are based on the air-freezing index.

AIR FREEZING INDEX (°F.-days)[2]	VERTICAL INSULATION R-VALUE[3,4]	HORIZONTAL INSULATION R-VALUE[3,5]		HORIZONTAL INSULATION DIMENSIONS PER FIGURE 403.3 (inches)		
		along walls	at corners	A	B	C
1,500 or less	4.5	NR	NR	NR	NR	NR
2,000	5.6	NR	NR	NR	NR	NR
2,500	6.7	1.7	4.9	12	24	40
3,000	7.8	6.5	8.6	12	24	40
3,500	9.0	8.0	11.2	24	30	60
4,000	10.1	10.5	13.1	24	36	60

For SI: 1 inch = 25.4 mm, °F. = 1.8°C. + 32.

1 Insulation requirements are for protection against frost damage in heated buildings. Greater values may be required to meet energy conservation standards. Interpolation between values is permissible.

2 See Figure 403.3b for Air Freezing Index values.

3 Insulation materials shall provide the stated minimum R-values under long-term exposure to moist, below-ground conditions in freezing climates. The following R-values shall be used to determine insulation thicknesses required for this application: Type II expanded polystyrene—2.4R per inch; Type IV extruded polystyrene—4.5R per inch; Type VI extruded polystyrene—4.5R per inch; Type IX expanded polystyrene—3.2R per inch; Type X extruded polystyrene—4.5R per inch. NR indicates that insulation is not required.

4 Vertical insulation shall be expanded polystyrene insulation or extruded polystyrene insulation.

5 Horizontal insulation shall be extruded polystyrene insulation.

Source: Council of American Building Officials (95) Table 403.3.

15

Estimating the Job

When estimating any project, the materials and labor are based on the expertise and experience of the workers. The better the workers, the faster and more efficiently they will perform their job. To develop accurate estimates, compare the actual materials used and time spent on each phase of the job with the estimated quantities of material and labor. Use the results of the comparisons to fine-tune estimates for future projects.

Estimating Site Preparation Costs

Estimating for any project should start with the cost of the plans. Whether the plans come from an architect, designer, plan service, or are created in house, the cost to prepare the plans and specifications must be included in the estimate. If the plans are purchased from a plan service, the plot plan indicating the placement of the structure on the lot must be created. Include the cost of this extra drawing in the estimate.

Permit costs and local fees also must be added to the estimate. These costs vary from one municipality to another. In some localities these costs are minimal, while in other communities the costs can be tens of thousands of dollars. Municipalities with high fees sometimes use the permit cost to limit the size and cost of dwellings.

Many of the fees are basic and easy to understand. They include basic building permits and *tap fees*, which are the fees charged to tap into the municipal water and sewer system. Some municipalities include *impact fees* based on the affect new construction has on the community, such as the potential impact of new residences on the school system. This charge might be based on the square footage or the number of bedrooms in the home.

Building fees vary from one community to the next. Verify the fees that apply to the residential project and include them in the estimate. Some examples of building fees are:

- Inspections
- Footings
- Foundation insulation
- Framing
- Plumbing
- Heating
- Electrical
- Wall insulation
- Drywall
- Ceiling insulation
- Final inspections
- Connection charges
- Water meter charge
- Water tap charges
- Sewer tap charges
- Curb-cut permits
- Occupancy permits

Just because municipalities don't have public water and sewer systems does not mean that they won't charge these fees for future services. In addition to the cost of installing a private well and septic system, tap fees may need to be included in the estimate.

Next estimate the costs of laying out and clearing the lot. If the lot corners are not staked, include the cost of a pin survey. After finding the corners, two workers usually can lay out a simple residence in two hours. Additional time is required for a more complex residence.

The costs associated with clearing the lot can vary greatly. During the construction of most residential developments, lots usually are cleared of most of the vegetation and replaced with an annual grass for erosion control. If this is the case, a tractor with a box blade can remove and stockpile the topsoil and build a level pad in a few hours.

If the site is heavily wooded, the cost to clear the lot will increase significantly. An experienced equipment operator can push over a tree 12 inches in diameter in about 10 minutes. The time doubles to 20 minutes for trees up to 24 inches in diameter and triples to 30 minutes for trees up to 36 inches in diameter. In addition, it takes a crew of two workers an hour or two with a chainsaw to cut up trees up to 12 inches in diameter into 24-inch sections and chip the branches into mulch. The time doubles for trees up to 24 inches in diameter and triples for trees up to 36 inches in diameter.

Trees usually can be pushed from the site and the tractor can build the pad while the trees are being cut up. In rural areas, burning may be an alternative. Usually a burning permit is required and the burning is limited to days with low fire danger.

The cost of excavating a basement depends on the topography of the site and the depth of the basement. The height of the basement walls and the height of the basement above the finished grade determine the depth of the excavation. Basement walls usually are 8 feet high with an 8-inch-thick footing and the top of the foundation usually is 8 inches above grade. These dimensions require the excavation to be 8 feet deep at its lowest point. If the lot has a slope to it, the excavation will be less at the other corners of the lot.

If the building site has a uniform slope with no sudden drops or high points, determine the depth at each corner. Average these depths and multiply the average depth by the footprint of the excavation. Remember that an additional 2 feet of over-dig is required to provide working room. This calculation should give the volume of soil that needs to be excavated in cubic feet. Divide this quantity by 27 to convert to cubic yards, the common measurement unit for excavation quantities.

The time needed to excavate this amount of soil is based on the capacity of the equipment. A backhoe with a 1-cubic-yard bucket can excavate 90 cubic yards of loose sandy loam per hour. With more common clay, this quantity is reduced to 70 cubic yards. If rock is present, after it has been blasted only 50 cubic yards can be excavated in an hour. This does not include the time it takes to dump the soil into trucks and remove it from the site.

The amount of soil that must be moved from the sight is more than what is dug from the hole in the ground because the soil is no longer compacted. It is looser and swollen, so the pile is bigger. Each soil type has different swell factors. Typical swell factors are:

- Clay—120 percent
- Gravel—115 percent
- Silt—120 percent
- Sand—115 percent
- Topsoil—130 percent

Estimating removal costs depends on the size and number of trucks, and the distance the soil must be moved. The time it takes for a truck to be loaded, driven to the dumpsite, emptied, and returned to be refilled is the *cycle time*. In many cases in residential construction the operator of the excavator is also the driver of the truck. In some cases several trucks can be used. If this is the case the excavator can dig the basement while the trucks are away from the site.

The following is a sample estimate based on the plan for the residence shown in Figure 15-1. This is the foundation plan for a 48×28-foot residence with a 24×24-foot attached garage. This structure will be built with 8-foot-tall foundation walls for the basement and a stemwall for the garage. The garage will not be excavated. An additional 2 feet of over-dig will increase the size of the bottom of the excavation to 52×32 feet. This means that the area of the excavation is 1,664 square feet.

The front corner (corner A) near the garage is the highest point of the building site. The depth at this point is 8 feet. The plot plan indicates corner B has an elevation about 2 feet lower than corner A. The depth of the excavation at this point is 6 feet. Corner C is 5 feet lower than Corner A so the depth of the excavation is 3 feet. The last corner is approximately 4 1/2 feet lower than Corner A so the excavation at this point is 3 1/2 feet.

Adding together the depths of each corner (8 + 6 + 3 + 3.5 = 20.5 feet) and dividing by 4 shows that the average depth of the excavation is 5.125 feet. Multiplying the area of the foundation by the average depth (1,664 square feet × 5.125 feet) calculates that the volume of soil to be excavated is 8,528 cubic feet. Dividing this volume by 27 converts the quantity to 315.85 cubic yards, or rounded approximately 315 cubic yards.

Most of the soil is clay, so 70 cubic feet of soil can be excavated in an hour. If a 1-cubic-yard bucket is used, it would take 4 1/2 hours to dig the basement. The top 6 inches are topsoil, which can stay at the site (1,664 square feet × 1/2 foot/27 = 31 cubic yards). Subtracting this from the total volume to be excavated (315 cubic yards-31 cubic yards) determines that 284 cubic yards of clay must be removed from the site.

Because clay swells, the 284 cubic yards must be increased by the swell factor (284 cubic yards × 120 percent) to determine the actual volume of soil that must be removed. The result is that 341 cubic yards of soil need to be removed. If the cycle time to load, haul, and dump each 20-ton truck capable of hauling 15 cubic yards per load is 30 minutes, it will take 11 hours for one truck to move the entire amount. Adding additional trucks can lessen the time. Because only one loader will be used, the maximum number of trucks depends on how long it takes to position and load each truck.

The exact cost of the site work can be calculated by multiplying the quantities by the appropriate unit costs for the region. A drawing for excavation quantities is shown in 15-2.

Estimating Concrete Footings

The quantity of material needed to place either concrete or gravel footings is determined by multiplying the width, depth, and length of the footings. If the footings are formed, add the dimensional lumber for the forms, stakes, spreaders, and nails to the amount of concrete and rebar needed to cast the footings.

FIGURE 15-1. Estimates are calculated on this drawing of a residential foundation.

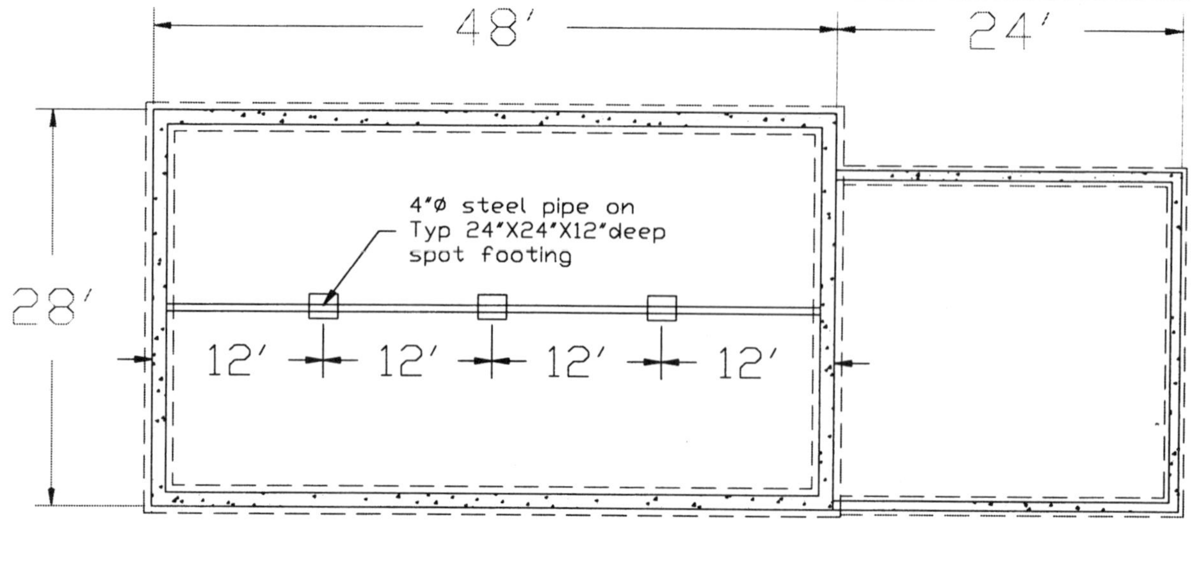

FIGURE 15-2. Excavation quantities are illustrated in this drawing.

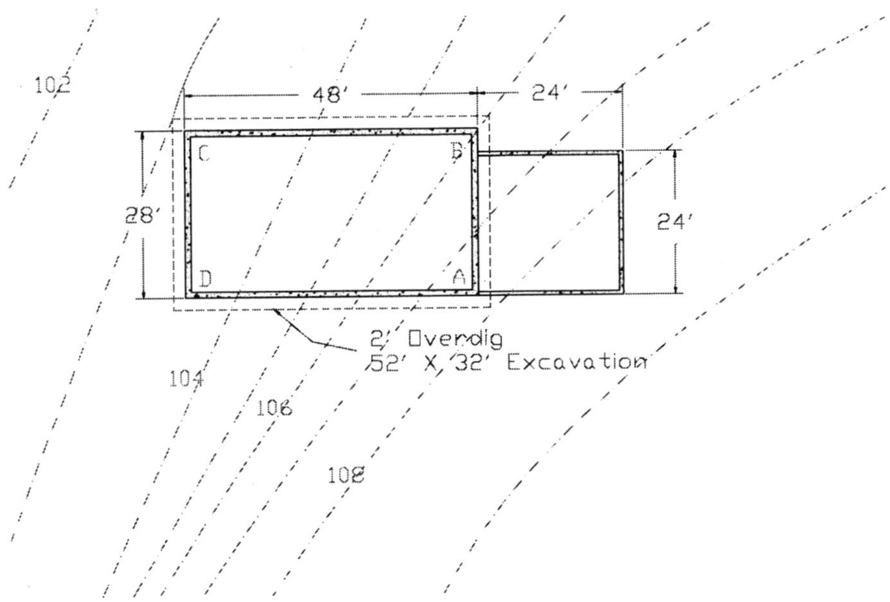

To determine material quantities start by converting each take-off dimension into feet. Then add the length of the footings together. If the footings have different widths or depths add up the dimensions of like footings and keep the totals separate. For example, a house may have 20 feet of 8-inch-deep footers that are 16 inches wide and 35 feet of 12-inch-deep footers that are 24 inches wide.

Calculating Form Materials

To find the amount of dimensional lumber required for the forms, add together the total length of each footing of an equal depth and double it. If the forms are 8 inches deep, use 2×8s, for 10-inch-thick footings use 2×10s, and so on. Convert the linear feet into board lengths. Usually 12- and 14-foot lengths are the cheapest per foot, so split the order into these lengths. Order a few extra feet to be cut up and used for stakes or bracing.

A stake is needed for every 3 feet of lumber ordered. To determine the number of spreaders needed, divide the length of the footing by 4. Many contractors like the convenience of having the materials precut so they use additional stakes for spreaders. If 1× material is used, multiply the width of the footing in feet by the number of spreaders needed to get the linear board feet.

Don't forget to include duplex nails in the estimate. Order one pound of 8-penny duplex nails for every 25 stakes and spreaders and one pound of 16-penny duplex nails per 100 feet of footing.

The last item needed for the forms is the releasing agent. Usually, a gallon of releasing agent will cover 400 linear feet of form.

Determining Labor Costs

Use the linear footage of the forms to calculate labor costs. Three-person crews can lay out and erect about 25 feet of footing forms in an hour. This same crew can place the concrete for about 150 linear feet of footing in an hour.

The amount of labor needed to dig a footing is dependent on the method used. Trenchers and backhoes usually can dig about 15 cubic yards of soil in an hour. A worker generally can clean the loose soil from 200 feet of trench per hour.

Calculating the Cost of Reinforcement Materials

If steel reinforcement is required, calculate the amount using the length of the footing. Usually two runs of rebar are placed in a footing. Check the foundation detail to verify this specification. Multiply the number of runs of rebar by the total length of the footing. Increase this amount by 5 percent to allow for the 2-foot overlap of the rebar sections. Then divide the total amount by 20 feet and round up to the nearest whole number to get the number of 20-foot lengths of rebar needed. Order an 18-inch steel pin or rebar chair for every 4 feet of footing. A crew of three usually can place and tie together about 300 feet of rebar per hour.

Determining Concrete Amounts

The method for determining the amount of concrete required is the same for both a formed and trenched footing. Change each measurement into feet with partial foot measurements converted to decimals to lessen the likelihood of making an error. Multiply the length, width, and depth of each footing to get the quantity of concrete required in cubic feet. Then divide this amount by 27 to convert the quantity to cubic yards.

Estimating Spot Footings

The material for spot footings also must be calculated. Spot footings for posts and a fireplace will be described on the foundation plan, usually with the dimensions for each of these items. Convert each measurement into feet and then multiply the length by the width and then by the depth. Then multiple this figure by the number of spot footings of each size. Divide the total quantity required for all spot footings by 27. Add the resulting number of cubic yards of concrete to the amount needed for the footings.

Estimating Gravel Footings

Gravel footings usually are placed in a trench. Convert the dimensions to feet with partial foot measurements converted to decimals. Multiply together the length, width, and depth in feet of the footings and divide by 27 to calculate the number of cubic yards of gravel needed. Then multiply the number of cubic yards by 1.35 to convert the amount to tons, which is the unit usually used to order gravel. The labor to excavate the footing and place the gravel is calculated using the same factors required for estimating a concrete footing.

Estimating Footing Blocks

Estimates for the quantity of footing blocks are based on the length of the footing. Divide the total length of the footing by three if 4×16-inch stretcher units are used and by four for 3×20-inch units. Then increase the quantity by 5 percent for waste. A bag of Type M masonry cement and 1/4 cubic yard of mason's sand is needed for every 40 units. A crew of three can place about 50 units per hour.

Estimating Materials for Block Foundations

To estimate the materials needed for block foundations, determine the length, height, and thickness of the foundation walls from the foundation and section plans. Some foundation plans provide a written indication of the height, such as 12-course walls. To calculate the number of blocks needed for each course multiply the length of each wall by .75 and round up to the next whole number. Then multiply this quantity by the number of courses. The number of courses can be found by counting them on the section plan or by multiplying the height of the wall by 1.5 and rounding up, if necessary, to the nearest whole number.

Add two sash blocks for each 16-inch-tall window, three for each 24-inch-tall window. Eight sash blocks are needed for each door. A pair of 4-inch-wide lintels is needed to bridge between foundation walls of differing heights. Usually 48-inch-long lintels will suffice. Longer lintels may be needed as the distance between the footings for each wall increases.

Once the total number of blocks is calculated use the quantity to figure the amount of masonry cement and sand needed. Most masons can lay 28 blocks with one bag of masonry cement. It also takes 1/10 of a yard of sand for 28 blocks. This amount of sand does not account for the sand at the bottom of the pile that is contaminated with dirt so add an extra yard of mason's sand to compensate for this.

Each pilaster requires a section of rebar equal to the height of the wall. The rebar is held in place with grout. The grout usually is a concrete mixture of portland cement, course sand, and 3/8-inch gravel. It takes approximately 0.25 cubic feet of grout to fill the cores of an 8-inch block, 0.33 cubic feet for a 10-inch block, and 0.45 cubic feet for a 12-inch block. In addition, one framing anchor usually is placed every 6 feet.

Before placing the backfill, temporarily brace the walls every 8 feet by placing a 7- to 8-foot-long 2×6 vertically against the wall. Hold it in place with a 10-foot-long 2×4 with one end attached diagonally about 6 feet up the 2×6 and the other end attached to a 24-inch-long stake driven into the ground. Hold the brace together with 16d duplex nails.

Most blocks are laid using a crew of workers consisting of several masons and a tender. The skilled masons set up the work and lay the blocks while the tender, sometimes referred to as the laborer, supplies the masons with blocks and mortar. The tender also is responsible for mixing the mortar. A crew of three masons and one tender usually can lay between 600 and 700 8-inch lightweight blocks per day. The quantity decreases by 10 percent when heavier and wider blocks are used. This estimate includes the time required to set up the job, build pilasters, place a parge coat on the exterior side of the block wall, and clean up.

Many masonry contractors get paid by the block. For a specific per-block price the mason will furnish all of the labor required to erect the foundation. The price may be supplemented by charges for items such as doors, fireplaces, and bay windows that require more time and skill. The mason also may modify the per-block charge based on the weight and size of the block, charging more for the larger or heavier units.

Estimating Insulating Concrete Forms

Estimates for foundations built with insulating concrete forms (ICFs) must take into account the specifications of the system used. One method that can be adapted for all systems is to calculate the area of each of the foundation walls by multiplying each wall's length by its height. Round up the height to a distance equal to an increment of the height of the ICF being used. Once the total area is calculated divide it by the area of one side of the ICF. The quotient will be the quantity of ICFs needed. Although most ICFs are difficult to damage, adding a few extra will ensure that a sufficient number are delivered to the jobsite.

When estimating for the panel system, remember that a form may be needed for each side when figuring the quantities needed. Calculate the number of wall ties based on the number of pairs of panels. Each system requires a specific number of ties to hold together each pair of panels.

The sizes and quantities of the required rebar are determined by the site conditions. The quantity of concrete is again unique to each system. Multiply the amount required for each ICF by the number of units needed. Because little waste exists with this system and a few extra ICFs were included in the estimate, do not add concrete for waste.

There is a definite learning curve when erecting ICFs. After several jobs, however, a four-person crew can set the forms on a previously placed footing for a 26×30-foot residence in one day. The same crew will need three hours to place the concrete into the forms and another hour to align the forms.

Estimating Cast-in-Place Walls

Forms are required to cast walls. Manufactured forms are available for rent or purchase. Plywood and dimensional lumber also can be used to build forms at the jobsite. Manufactured forms are available in numerous widths and heights. Usually a number of the widest forms are chosen and the remaining void is filled with smaller units. Some systems use corner units to speed up the installation process. Companies that provide the forms generally request a copy of the foundation plan and figure the purchase price or rental charge for the units needed. They also provide bracing and stakes for the project.

Plywood for site-built forms can be calculated from the inside and outside perimeter measurements of the foundation walls. Most builders use 4×8-foot sheets placed vertically for walls taller that 4 feet and horizontally for 4-foot-tall walls. The sheets may be ripped to 2×8-foot sheets, although builders try to limit the number of plywood sheets that are ripped. Most of these forms are reused.

Dimensional lumber also is needed for walers and bracing. One pair of boards is placed along the footing and four boards are needed for every 16 inches of wall height. The length and quantity of this lumber can be calculated by multiplying the wall length by the number of walers needed.

Wall ties are required for both manufactured and site-built forms. Site-built plywood forms use approximately one tie for every 3 square feet of plywood. Each manufactured system requires a specific amount of ties.

In addition, a releasing agent is needed to make it easier to remove the forms and to increase the longevity of the forms. The coverage per gallon is stated on the container, but most releasing agents cover about 400 linear feet. Remember that the forms on both sides of the wall need to be coated.

Concrete is the primary material for cast-in-place walls. The quantity is based on the height, length, and thickness of the foundation walls. Because the walls are placed on footings use those calculations for the foundation wall lengths. The walls usually are of differing heights. For example, there may be shorter walls around the garage. Separate the walls by length and thickness. Multiply the length of each wall by its height and thickness, remembering to keep all units the same. Convert the 8-inch-thick wall to 0.67 feet so the resulting figure will be in cubic feet. Divide this by 27 to convert the number to cubic yards and then round it up to a 1/4-yard increment, e.g., 3 1/4 yards instead of 3 1/8 yards.

The quantity of reinforcing steel needed depends on site and soil conditions. To calculate the amount of rebar required, multiply the number of horizontal runs required by the length of the walls. Add 5 percent extra to compensate for areas where the rebar overlaps. Divide the total length by the length of rebar available. Most suppliers carry rebar in 20-foot lengths, but longer and shorter pieces may be available. The number of vertical pieces can be calculated by dividing the length of the foundation walls by the required on-center spacing. Generally, it is more economical to order vertical pieces in lengths equal to the height of the walls. Some contractors, however, prefer to order longer units, bend them at the top, and then tie them into the horizontal rebar.

The costs for most elements required for foundation walls, including labor costs, are calculated on a per-square-foot basis. A crew of three carpenters and a laborer can erect about 1,000 square feet of plywood forms per day. A two-person crew usually can place the concrete for the foundation walls of a 1,500-square-foot residence in four hours. One laborer can remove about 1,000 square feet of forms in an eight-hour day.

Manufactured forms can be erected faster that site-built forms. A carpenter can set about 600 square feet of forms in a day. It takes about the same time to cast the wall. The forms can be stripped and cleaned at a rate of about 2,400 square feet per day.

Estimating Slab Materials

Concrete slabs are placed on a compacted subgrade. The depth of the granular material in the subgrade is equal to the height of the stemwall minus 4 inches. If the pad is not level and the depth is not constant, calculate an average depth. Multiply the depth in feet by the square footage of the floor area. Divide the total by 27 to get the number of yards of fill needed. Convert the number of yards to tons by multiplying by 1.35.

The quantity of 6-mil polyethylene needed for the vapor barrier can be determined by dividing the coverage of a roll by the square footage of the slab. Allow for extra material because the sheets must overlap by at least 12 inches. Be sure to include tape to join the sheets and repair any breaks.

The amount of concrete can be calculated by multiplying the thickness of the slab in feet by the square footage of the slab. If the slab is thicker under load-bearing walls, use the length, thickness, and depth of this area to calculate the additional amount of concrete.

The amount of time required to spread the subgrade depends on how close the materials can be delivered to the slab's location and the machinery used. Two individuals can take two days to spread the fill under a 1,500 square foot residence with shovels and wheelbarrows. The time can be reduced to less that three hours if the same two-person crew uses a skid-steer loader. An individual operating a conveyor mounted on a truck, shown in Figure 15-3, can complete the task in about 30 minutes.

Two individuals can place about 1,000 square feet of vapor barrier in an hour.

A four-person crew can place, screed, and finish about 250 square feet of concrete per hour. Because the concrete's set time is dependent on temperature, slump, and the use of admixtures this time can fluctuate.

Estimating Frost-Protected Foundations

The basic materials required for this foundation system are calculated in the same manner as for

FIGURE 15-3. A conveyor system can reduce the hours required to spread the subgrade.

traditional foundations. Add the volume of additional gravel under the slab and around the foundation. To find the quantity of materials needed to insulate and drain the foundation multiply the length of the footing by the height of the footing plus the distance the insulation must extend beyond the perimeter of the foundation. Remember that this distance is dependent on local climatic conditions. A crew of two people usually can spread about 6 tons of gravel per hour and place the insulation at a rate of 400 square feet per hour.

For additional estimating resources for complete residential construction, consult *Estimating Home Construction Costs* by Jerry Householder, 1999, and *Action Estimating* by William Avon Privette, 1998. Both publications are available from Home Builder Press.

Chapter Checklist

When estimating residential foundation construction costs, consider these questions.

- How does this job compare to the final costs of other, similar jobs?
- Have all dimensions on the plans and sections been verified?
- Are there site conditions, such as poor access, that can hinder the progress of the job and add to the cost?
- Will the task be done in-house or are subcontractors needed?
- When renting equipment or material, is delivery included in the cost?

Appendix

One-Call Directory

Alabama

Alabama Line Location Center 1-800-292-8525

Alaska

Locate Call Center of Alaska, Inc. 1-907-278-3121

Arizona

Arizona Blue Stake, Inc. 1-800-782-5348

Arkansas

Arkansas One Call System, Inc. 1-800-482-8998

California

Underground Service Alert North 1-800-227-2600
Underground Service Alert South 1-800-227-2600

Colorado

Utility Notification Center of Colorado 1-800-922-1987

Connecticut

Call Before You Dig 1-800-922-4455

Delaware

Miss Utility of Delmarva 1-800-282-8555

Florida

Sunshine State One Call of Florida, Inc. 1-800-4324770

Georgia

Utilities Protection Center, Inc.
1-800-282-7411

Hawaii

Hawaii One-Call 1-800-282-7411

Idaho

Dig Line 1-800-342-1585
Palouse Empire Underground Coordination Council 1-800-822-1974
Kootenai County Utility Coordination Council 1-800-428-4950
Utilities Underground Location Center 1-800-424-5555
One Call Concepts—Idaho 1-800-626-4950
Shoshone County One Call 1-800-398-3285

Illinois

Julie, Inc. 1-800-892-0123
Chicago: Digger 1-312-744-7000

Indiana

IN Underground Plant Protection Services, Inc.
1-800-382-5544

Iowa

Underground Plant Location Service, Inc.
1-800-292-8989

Kansas

Kansas One Call Center 1-800-DIG-SAFE

Kentucky

Kentucky Underground Protection, Inc.
1-800-752-6007

Louisiana

Dottie System, Inc. 1-800-272-3020

Maine

Dig Safe—Maine 1-888-344-7233

Maryland

Miss Utility of Delmarva 1-800-282-8555
Miss Utility 1-800-257-7777

Massachusetts

Dig Safe—Massachusetts 1-888-344-7233

Michigan

Miss Dig Utility Comm. System 1-800-482-7171

Minnesota

Gopher State One Call 1-800-252-1166

Mississippi

Mississippi One Call System, Inc.
1-800-227-6477

Missouri

Missouri One Call System, Inc.
1-800-344-7483

Montana

Utilities Underground Loc. Center
1-800-424-5555
Montana One Call 1-800-551-8344

Nebraska

Diggers Hotline of Nebraska 1-800-331-5666

Nevada

Underground Service Alert of NV
1-800-227-2600

New Hampshire

Dig Safe—New Hampshire 1-888-344-7233

New Jersey

Garden State Underground Plant Location Service 1-800-272-1000

New Mexico

New Mexico One Call System, Inc.
1-800-321-ALERT

New York

Underground Facility Protection Organization
1-800-962-7962
New York City—Long Island One Call Center
1-800-272-4480

North Carolina

North Carolina One-Call Center
1-800-632-4949

North Dakota

Utilities Underground Location Center
1-800-454-5555

Ohio

Ohio Utilities Protection Service 1-800-362-2764

Oklahoma

Call Okie 1-800-522-6543

Oregon

Utility's Underground Location Center 1-800-424-5555
Douglas Utility's Coordinating Council 1-503-673-6676
Josephine Utility's Coordinating Council 1-503-476-6676
Utility's Notification Center 1-800-332-2344
Rogue Basin Utility Coordinating Council 1-503-779-6676

Pennsylvania

Pennsylvania One Call System, Inc. 1-800-242-1776

Rhode Island

Dig Safe—Rhode Island 1-888-344-7233

South Carolina

Palmetto Utility Protection Service Inc. 1-800-922-0983

South Dakota

South Dakota One Call 1-800-781-7474

Tennessee

Tennessee One Call System 1-800-351-1111

Texas

Lone Star Notification Center 1-800-669-8344
Texas Excavation Safety System 1-800-344-8377
Texas One Call System 1-800-245-4545

Utah

Blue Stakes Location Center 1-800-662-4111

Vermont

Dig Safe—Vermont 1-888-344-7233

Virginia

Miss Utility of Virginia 1-800-552-7001
Miss Utility 1-800-257-7777
Miss Utility of Delmarva 1-800-441-8355

Washington

Utilities Notification Center 1-800-332-2344
Grays Harbor & Pacific County Utility Coordinating Council 1-206-532-3550
Utilities Underground Location Center 1-800-424-5555
Chelan-Douglas Utility Coordinating Council 1-509-663-6111
Upper Yakima County Underground Utilities Council 1-800-553-4344
Inland Empire Utility Coordinating Council 1-509-456-8000
Utilities Council of Cowlitz County 1-360-425-2506
Palouse Empire Utility Cord Council 1-800-822-1974

Washington, District of Columbia

Miss Utility 1-800-257-7777

West Virginia

Miss Utility of West Virginia, Inc. 1-800-245-4848

Wisconsin

Diggers Hotline, Inc. 1-800-242-8511

Wyoming

Wyoming One-Call 1-800-348-1030
West Park Utility Coordinating Council 1-307-587-4800
Converse County Coordinating Council 1-800-562-5561
Fremont County Utility Coordinating Council 1-800-489-8023
Central Wyoming Utilities Coordinating Council 1-800-759-8035
Albany County Utility Coordinating Council 1-307-742-3615
Southeast WY Utility Coordinating Council 1-307-638-6666
Southwest Wyoming One Call 1-307-362-8888
Utilities Underground Location Center 1-800-454-5555
Carbon County Utility Coordinating Council 1-307-324-6666
Call-In Dig-In Safety Council 1-307-682-9811

Glossary

A

Accelerators—An admixture such as calcium chloride that speeds up the hydration of concrete. Usually used when the ambient temperature is below 50° F.

Acrylic Coatings—Latex-based materials applied to the exterior of a foundation wall during the damp-proofing or waterproofing process.

Admixtures—Ingredients other than water, aggregates, and cements that influence the physical properties of the concrete. (See *Accelerators, Calcium Chloride.*)

Aggregate—Natural sand, gravel, and crushed stone in the cement mixture.

Air Entrainment—The intentional incorporation of minute air bubbles in concrete to improve its durability and workability.

Air-Freezing Index—The cumulative total of average annual days when the temperature is below 32° F. It is a measurement of the extent of frigid air temperature for a geographic region.

Alpha Track Detector—A device used to detect the presence of radon in a residence.

Ammonical Copper Arsenite—A chemical used to pressure-treat preserved wood.

Anchors—Metal devices used to attach structural members, typically sill plates, to foundations.

Anchor Bolts—A threaded steel rod used to connect structural members to foundations.

Annealed—The process of heating metal and then cooling it slowly to prevent brittleness.

Approved—A material or method that is accepted by a building code authority.

Angle of Layback—The angle at which a mound of granular material such as dirt, sand, or gravel is stable.

Applied Rubber Polymer Sprays—A polymer-based material sprayed on the exterior of a foundation wall for damp-proofing or waterproofing.

Aquifer—A plentiful underground water source.

Asphaltic Coatings—Bituminous, petroleum- or coal-tar-based materials applied to the exterior of a foundation wall to damp-proof or waterproof the surface.

B

Backfill—The soil that is placed around the foundation walls and compacted to compensate for the over excavation needed to construct the footings and walls.

Basement—Space surrounded by foundation walls that has the potential to be used as habitable space; must have a minimum clearance of 7 feet 6 inches.

Batterboards—Horizontal boards attached to stakes placed 4 to 6 feet outside of the corners where foundation walls intersect. Nails or marks are placed on the horizontal boards to provide semipermanent reference points to define the exterior of the foundation walls.

Beam Blocks—Concrete masonry units designed to create a horizontal void when laid in a course in a wall. A reinforcing steel dowel encased in concrete is placed in the void.

Bearing Capacity—The amount of load or weight that a soil can support.

Bell Piers—A concrete cylinder that flairs out at its base to provide a larger area of support. The base of the flaired portion should be at least a 6-inch-thick cylinder. The concrete in the cylinder may or may not be reinforced with deformed steel.

Bentonite—A highly expansive clay used for waterproofing.

Blocking—see *Blockouts.*

Blockouts—A temporary device placed in a foundation wall during construction to fill the space where an opening for a door or window is specified. Blockouts can be made from lumber, plywood, styrofoam, or any other material that can be secured in a stable position.

Blowout—A term that refers to the failure of a forming system during the casting of a foundation wall. In a blowout concrete escapes from the forms.

BOCA (Building Officials and Code Administrators)—An organization that produces model building codes.

Bonding Agents—Chemicals that promote the adhesion of one layer of concrete to another that has been cast previously.

Bond Beams—Continuous horizontal members of reinforced concrete in a wall designed to unify the structure and distribute the load imposed on the wall.

Box Blades—A device attached to the rear of a tractor that is used to move dirt; commonly used to level a building site.

Bracket Clamps—A device used to secure walers to wall ties used in the construction of cast-in-place concrete foundation walls.

Brick Ledge—An extrusion built into the foundation or footing to support a masonry exterior on the side of the residence.

Buck—Wood, plastic, or another material set in a foundation wall opening to facilitate the fastening of a window or door frame to the wall.

Builder's Level—A device that uses a telescope fixed on a tripod to determine points of common elevation.

Building Paper—An asphalt-impregnated paper used in construction, commonly referred to as felt paper. It is available in rolls 18 or 36 inches wide and in weights of 15 or 30 pounds per 100 square feet.

Bull Float—A T-shaped tool with a large flat blade attached with a hinged joint to a long handle used to smooth the surface of freshly screeded concrete flatwork.

C

CABO (Council of American Building Officials)—An organization that produces a model building code.

Cage—An array of vertical reinforcing steel tied to circular hoops. Commonly used in concrete piers.

Calcium Chloride—A salt sometimes supplied in solution or found as an ingredient in an admixture used to accelerate the setting or strength gain of freshly placed concrete. May contribute to rusting of reinforcing steel or discoloration of concrete surfaces.

Caulking—The process of filling a joint with a material suitable for sealing out dirt and moisture. Also, the sealing material. Better-grade materials used for this purpose commonly are known as sealants.

Cells—See *Cores.*

Cement—A powder that interacts with water to form a bonding agent. Different types of cement use various materials depending on the intended use. (See *Masonry Cement, Portland Cement.*)

Cementitious—A mixture of cement and another natural ingredient, such as a grout made of portland cement and coarse sand.

Chairs—Metal or plastic objects designed to hold reinforcing steel or welded wire fabric in the proper position during the casting of concrete flatwork.

Chamfer Strips—A piece of material, commonly wood, used to form control joints or angled corners in cast-in-place concrete walls.

Charcoal Monitor—A device used to detect the presence of radon gas in a residence.

Chase—A void created in a wall or floor into which plumbing, mechanical, or electrical components are placed.

Chromium Copper Arsenic (CCA)—The most common chemical used in the process of pressure treating preserved wood.

Chromated Zinc Chloride—A chemical used in the process of pressure treating preserved wood.

Clay Soils—Soils that are predominately made up of particles of decomposed rock smaller than .005 mm. Clay soils are sticky and moldable when wet and extremely hard when dry, making them poor soils on which to erect a structure. The volume of some clay soils changes dramatically when the moisture content changes.

Cohesionless—The lack of the physical or chemical attraction between soil particles; soil particles, commonly sand and gravel soils, that do not hold together.

Cohesive—The physical and/or chemical attraction soil particles have for each other. Clays are an example of a cohesive soil.

Coil Ties—Two threaded inserts welded to the ends of small-diameter steel rods and threaded through the ends of forms. Ties allow the forms to be adjusted to various widths for concrete placement.

Cold Joints—A joint, usually visible, in a concrete wall or floor where fresh concrete has bonded imperfectly or not at all to a previously placed layer of concrete. Occurs when too much time elapses between the two placements.

Come-Along—A tool with a wide blade similar to a hoe on a wooden handle that is used to pull away excess concrete or push concrete in front of a screed or bull float to smooth the surface of freshly screeded concrete flatwork. (See *Bull Float.*)

Competent Person—An individual with training and knowledge that can recognize hazardous conditions and has the authority to control and stop work until the hazardous situation is corrected.

Compressive Strength—A material's ability to be squeezed without breaking, usually expressed in pounds per square inch (psi).

Concrete Masonry Units (CMUs)—Manufactured building materials commonly referred to as concrete blocks.

Concrete—A composite material made of portland cement or other hydraulic cement, aggregate, water, and sometimes admixtures that hardens when the cement chemically reacts with water.

Control Joints—A straight-line indention purposely placed in a concrete or masonry structure to localize cracks caused by the expansion and contraction created by changes in the moisture content due to drying or seasonal changes in temperature.

Cores—The voids created in concrete masonry units during the manufacturing process.

Corner Blocks—Concrete masonry blocks that have two flat-face surfaces and at least one flat-end surface. Corner blocks are produced in 10- and 12-inch sizes with one 8-inch side to allow the mason to keep the block's end joints on 8-inch increments.

Crawl Space—The area between the ground and the underside of the floor joists. The space must be a minimum of 18 inches from ground level to the underside of the floor joists.

Cube—The general manner in which masonry block is sold; a cube contains 108 blocks and typically includes one corner block for every three stretcher blocks. Stretchers are ordered separately. (See *Corner Blocks, Stretcher Block.*)

Curing—The procedure of retaining the proper moisture content in freshly placed concrete. Water must be present for the chemical reaction that gives concrete its strength to occur.

Curing Agents or Compounds—A chemical applied to the surface of freshly finished concrete to influence the retention of water in the concrete and promote curing. (See *Curing.*)

Course—The levels of block in a foundation wall or stemwall.

D

Damp-proofing Process—A water-repellent process usually applied to the outside of a building, such as a foundation wall, to protect against flowing and shallow standing water but not deep standing water. Examples are parge coats and bituminous coatings (See *Parge Coat, Waterproofing.*)

Darby—A long, straight, flat surface with an inclined handle used in the early stages to level freshly placed concrete after it has been screeded; darbied (v.) is the act of using a darby on fresh concrete. (See *Screeding.*)

Deflecting—The degree to which an object bends from the weight of the object itself or when a load is applied to it. (See *Flexural Strength, Plasticity.*)

Deformed Bar—A steel reinforcing bar with raised deformations on the surface that interlock with the surrounding concrete to provide additional strength. Commonly referred to as rebar or reinforcing steel. (See *Rebar.*)

Dewatering System—A system that places pipes in the ground to collect water and direct it away from areas where it might create problems.

Dogs—Wedges that are inserted between the tendon that is precast in a post-tensioned slab and the live-end anchor. (See *Post-Tensioned Slab.*)

Drainage Mat—Fibrous boards or blankets generally made of fiberglass or plastic that are laid against the foundation walls to draw water away from the walls and into a drainage system.

Drainage Tiles—Hollow round perforated drainage pipe made of either clay or smooth-rigid or flexible-corrugated plastic referred to in some regions of the country as tiles.

Dry Stacking—The process of erecting a block wall without mortar; using dry blocks. (See *Dry Blocks.*)

Dry Blocks—Concrete masonry blocks (CMUs) that interlock and are placed together without mortar; once in place, a thin coat of cementitious material reinforced with strands of fiberglass is added to bond the blocks together and protect against moisture infiltration.

Duplex Nails—Double-headed nails used in temporary construction. Commonly used to erect forms for the placement of concrete; sometimes referred to a scaffold nails.

E

Electric Ion Chamber—A device used to detect the presence of radon gas.

Environmental Protection Agency (EPA)—A U.S. agency responsible for maintaining the quality of the environment.

Equivalent Fluid Density—The weight of a volume of soil treated as a fluid to determine the lateral load placed on the foundation wall by the soil adjacent to the residence.

Expansive Soils—A family of soils found in various parts of the country that contain a large portion of highly plastic or moisture-sensitive clay. Because these soils are sensitive to water their volumes increase, or swell, or decrease, or shrink, with changes in their moisture content. (See *Clay Soils, Plasticity.*)

F

Face Surfaces—The side of a concrete masonry unit that is visible when it is laid in a wall.

FASWALL—A brand name of concrete masonry units that use expanded polystyrene as an aggregate.

Faulting—The vertical movement of a slab, footing, or other member adjacent to a joint or crack so that there is an abrupt change in the surface elevation from one side of the joint to the other.

Fill—See *Backfill.*

Filler—Objects designed to be placed under a grade beam prior to placing concrete. The filler, commonly made of cardboard or polystyrene foam, provides an area under the beam that allows for the expansion of the underlying soil. After casting the filler either dissolves or is left to decay. The remaining void allows soil under the grade beam a place to expand without exerting any unwanted forces on the grade beam. The cardboard fillers decay from elements in the environment while polystyrene fillers are dissolved with solvent.

Filter Fabrics—One of a family of plastic materials, called geotextiles, that is intended to intercept and hold the particles carried in suspension in water flowing or seeping toward a subsurface drainage pipe.

Finished Elevation—The horizontal position of the first floor.

Flat Wall—A solid cast-concrete wall of uniform thickness.

Flame Spread Rating—A number that expresses the rate of combustibility by the speed of burn across the surface of a material.

Flexural Strength—A material's ability to resist bending. (See *Deflecting, Plasticity.*)

Flood Fringe—The area adjacent to a body of water that can be inundated by water deposited by a rainstorm with the potential to occur once every 100 years.

Floodplain—The area covered by both the floodway and the flood fringe. (See *Flood Fringe, Floodway.*)

Floodway—The area occupied by a body of water when it is at its normal level.

Floor Joists—One of a series of parallel beams, usually 2 inches thick, used to support floor and ceiling loads, and supported in turn by larger beams, girders, or bearing walls.

Flourochrome Arsenate Phenol—A chemical used in the process of pressure treating preserved wood.

Footer—The portion of a foundation system that transmits the load of the structure to the bearing soil. In some areas this name is synonymous with footing. (See *Footing.*)

Footing—The portion of a foundation system that transmits the load of the structure to the bearing soil. In some areas this name is synonymous with footer. (See *Footer.*)

Footing (Footer) Drain—A system designed to remove water from the base of a foundation wall.

Footing (Footer) Plate—The bottom structural member of a preserved wood foundation that passes the load of the structure from the foundation to the footing. The plate is a 2× at least 2 inches wider than the studs used in the foundation.

Form Ties—The metal, fiberglass, or plastic elements of a formed cast concrete system that holds both sides of the forms together.

Foundation—The structural elements that transfer the load of the residence to the earth.

Freeze-Thaw Cycle—The repeated occurrence of moisture freezing and thawing in the soil. Since water expands when it freezes this phenomenon exerts pressures on solid materials that can deteriorate the structural integrity of the object.

French Drain—A surface or subsurface drainage trench or ditch that contains gravel or stone but no pipe. A filter fabric may surround the gravel as an envelope.

Friction—The load placed on a piling that passes to the soil into which the piling is driven.

Friction-Bearing Piling—A piling that uses friction to support the weight of the house by balancing the load against the pressure exerted on the piling by the soil into which it has been driven.

Frost-Protected Shallow Foundation—A footing (footer) system that does not place the base of the footing below the frost line but provides protection from the affects of the freeze-thaw cycle. (See *Freeze-Thaw Cycle, Frost Line.*)

Frost Line—The depth of frost penetration. This depth varies in different parts of the country.

Furring Strips—Pieces of wood or metal applied to a wall or other surface to even it out or to serve as a fastening base for finishing materials.

G

Gamma Radiation Detector—A device used to detect the presence of radon gas.

Geotechnical Engineer—A registered professional engineer that specializes in the relationship between structures and the earth.

Geotextile—A woven or nonwoven, porous cloth used for drainage, filtration, reinforcement, or separation. The term includes materials used for filter fabrics and silt fences.

Grade—The finished ground level.

Grade Beams—Reinforced concrete beams placed at ground level and supported by pilings or piers.

Gravel—Soil particles with a diameter between 2 mm and 76 mm.

Ground Spline Blocks—Concrete masonry units designed for use with surface bonding mortar. The blocks are sized large since no mortar joints are utilized. A groove is manufactured into the top and bottom surface of the blocks into which an alignment spline is inserted when the wall is erected. (See *Mortar.*)

Groundwater—The water under the surface of the ground (subsurface). The name may imply an underground origin, but most subsurface water originates on the surface.

Grout—A thin mortar used to fill in cracks or voids such as those in concrete masonry units; also a fine plaster for finishing surfaces.

H

Half Block—A concrete masonry unit that is 8 inches long by 8 inches tall of varying thickness.

Hazardous Soil Area—A district with soils that are expansive, compressive, shifting, or otherwise are an unstable base on which to erect a structure.

Header—A horizontal structural member that supports the load over an opening such as a door or window; also called a lintel. (See *Lintel.*)

Heat Tape—An electrical tape that can be wrapped around pipes in a crawl space or other area to protect the pipes from freezing.

Heaving—The lifting of a slab or other structural member by the expansion of the soil on which it is resting; commonly associated with the freezing of moisture in the soil.

Honeycomb—Voids left in concrete where cement and sand particles have not filled the spaces among the coarse aggregate particles.

Hydrostatic Pressure—Pressure exerted by a standing body of fluid that increases with depth; commonly exerted by moisture in backfill.

Horizontal Reinforcement—Steel reinforcement placed horizontally in cast-concrete walls to provide additional strength. (See *Rebar.*)

I

Impact Fee—A charge imposed by a municipality for the impact the structure has on the infrastructure of the community.

Impervious—The quality of being impenetrable; water cannot pass through an impervious fabric.

Infiltration—The penetration of water into the surface of soil, rock, or concrete, such as foundation walls or pavement.

Insulating Concrete Forms (ICFs)—A concrete forming system using stay-in-place forms constructed of foam plastic insulation or a combination of foam plastic and other materials.

Integral Footing—The part of the floor slab or foundation that acts as a footing to transmit the load of the residence to the bearing soil. An example is the thickened edge of a monolithic slab that may or may not contain reinforcing steel.

J

J Bolt—An anchor rod with threads on one end and a bend on the other end that is embedded into concrete to attach a framing member to the foundation. (See *Anchor Bolts.*)

Joists—A series of parallel beams, usually 2 inches thick, used to support floor and ceiling loads, and supported in turn by larger beams, girders, or bearing walls. (See *Floor Joists.*)

L

Lap Splice—The joint formed by extending reinforcing bars past each other a specified distance to allow the stress on the members to be transferred between two adjacent parallel members.

Lateral Strength—The ability to resist pressure exerted in a horizontal direction by freshly placed concrete in the forms or by hydrostatic pressure against the foundation wall.

Lift—A layer of freshly placed concrete placed to a predetermined depth and allowed to cure for a set time so that the concrete can gain an initial set before the next layer is placed. This reduces the lateral loads on the forms.

Lintel—A horizontal structural element of reinforced concrete located above an opening to distribute loads to the walls around the opening. A concrete masonry unit that spans the gap in a block wall. Units longer than 32 inches are reinforced with steel dowels.

Load—The weight of the structure, furnishings, and people inside the structure, and forces of nature such as rain, snow, and wind.

M

Manufactured Forms—Interlocking panels produced to form cast-in-place concrete walls or slabs.
Masonry Cement—The powder, when mixed with water and sand, that is used to bond materials such as bricks, blocks, and stones together. It is available in several strengths (Types N, S, and M) for use in different building features and conditions.
Modular Sizing—Standardized dimensions of construction materials that create efficient use without modifying shapes.
Modulus of Elasticity—A numerical value that indicates the ability of a material to resist bending. It is the ratio of force applied on the material to its deformation.
Monolithic Slab—A continuous slab that uses only construction joints.
Mortar—A material made of masonry cement and sand used to bond masonry materials, such as blocks, together.
Mud Rake—A tool with a wide blade similar to a hoe on a wooden handle; also called a come-along. (See *Come-Along*.)
Mud Slab—A 2- to 6-inch layer of concrete placed below a structural concrete floor or footing to provide additional support on soft soil; also called a mud mat.

N

Net Free Area—The actual space available, such as the gross area of a vent minus all space taken up by frame work, netting, or other obstructions.
Nominal Sizing—The dimensions of an object in whole numbers. These dimensions usually are slightly larger that the actual size.

O

Occupational Safety and Heath Administration (OSHA)—A division of the U.S. Department of Labor that is responsible for worker safety.
Off-Gassing—The process of a material emitting vapors after manufacturing. This process can occur for several months after the product is installed in a residence.
Offset Stakes—See *Batterboards*.
Outfall Line—A solid pipe used to remove water collected by a footing drain away from the residence.
Oxidation—The interaction of a material with oxygen; within a reinforced foundation or slab oxidation can cause rebar to rust. (See *Rebar*, *Reinforced Slab*.)

P

Pad—A level building site cleared of vegetation.
Parge Coat—A thin coat of plaster or mortar on a vertical masonry surface; commonly applied to the inside of a wall for waterproofing.
Pencil Rods—Heavy steel wires that remain in the concrete.
Perched Water Table—A layer of groundwater above the true level of the water table. (See *Groundwater*, *Water Table*.)
Pier-and-Grade-Beam Foundation—A support system that transfers the load of a residence to horizontal members and then to the vertical members that finally transfers the load to stable soil. This foundation system can be used on sites that have stable soil or in situations where soils of suitable bearing capacity are overlaid with an undesirable soil.
Pier—A vertical support that provides bearing in the ground.
Pilaster—A projection from the face of the wall that extends the wall's full height to provide lateral support.
Piling—A steel, wood, or concrete member that is driven into the earth as part of a foundation system.
Placement Rate—The speed at which concrete is moved from the delivery vehicle into the forms.
Plasticity—The ability of an object to bend without breaking.
Plat of Survey—See *Plot Plan*.
Plate Shape—A characteristic of clay soil components that have flat, smooth top and bottom surfaces and thin edges; a thin steel member with a smooth, flat top and bottom.

Plot Plan—A dimensional drawing that shows the shape of the building site, the residence's position on the site, and other information such as sidewalks, driveways, easements, and utility locations.

Plyform—A specialized plywood used for concrete forms.

Plug—A device inserted into the delivery hose of a concrete pump to end the delivery of concrete and clean the hose.

Point-Bearing Piling—A piling that is driven to a predetermined depth, usually through unstable soil to a layer that can support the load of the residence.

Polyethylene Film—A 6-mil-thick sheet commonly used as a vapor barrier under concrete slabs.

Pond—Standing water within an excavated foundation or near the surface of a constructed foundation.

Portland Cement—The powder which, when mixed with water and aggregates, slowly reacts chemically with the water to form the bonding agent that holds the aggregate together, producing concrete.

Post-and-Beam Wall—An engineered foundation system created with insulating concrete forms (ICFs) that produces a wall with a horizontal reinforced concrete member supported on vertically reinforced concrete members placed more than 12 inches on center. (See *Cores, Insulating Concrete Forms.*)

Post-Tensioned Slab—A slab or other concrete structure in which steel cables are placed and then stretched after the concrete has set, putting the slab in tension. The technique makes best use of concrete's compressive strength and the steel's tensile strength. (See *Tendons.*)

Powder-driven Nailer—A tool that uses a gunpowder charge to drive specially designed nails into concrete or steel.

Preserved Lumber—See *Pressure Treated Wood.*

Pressure Pipe—A solid pipe through water collected in a sump is pumped to move the water away from the residence. Most local building codes prohibit this pipe from connecting to the sanitary sewer.

Pressure Treated Wood—Lumber and plywood that has been impregnated with chemicals using special facilities and chemicals. This process is used to alter the characteristics of the wood commonly making it resistant to decay or combustion.

Pounds Per Square Inch (psi)—An abbreviation for the units in which pressure is expressed.

Q

Quicksand—A soil layer saturated with water that has become fluid and lacks a bearing capacity. (See *Bearing Capacity.*)

R

Radium—A naturally occurring mineral found in soils and bedrock.

Radon Gas—A vapor created during the radioactive decay of radium, an element that is present to some degree in all soils; enclosed areas such as crawl spaces must be vented to prevent gas buildup.

Raft Foundation—A system that uses a reinforced monolithic slab and walls as the main support elements.

Raft Slab—A 2- to 6-inch layer of concrete below a structural concrete floor or footing to provide additional support on soft soil; also called a mud mat. (See *Mud Slab.*)

Raker—A diagonal brace that extends from the top of the form of a cast-in-place foundation wall to the interior of the formed space.

Ram—The weight or striking part of a piling driver; the plunger of a force pump, such as a sump pump.

Rebar—A steel reinforcing rod with raised deformations on the surface that interlock with the surrounding concrete.

Reinforced Slab—A slab designed by a structural engineer incorporating reinforcing steel to increase the slab's resistance to the flexural forces exerted by the underlying soil. Applying stress to a reinforced slab makes the slab stronger.

Releasing Agent—A chemical applied to forms that prevents the bonding of concrete to the form. Motor oil is not an acceptable releasing agent.

Retention Rate—The concentration of chemicals retained in a given batch of pressure-treated wood stated in terms of pounds per cubic feet (pcf). The retention rate is stamped on pressure-treated wood; the rate must comply with established standards for the system in which the wood will be used.

Rock—A solid mineral that must be blasted before it can be removed.

Rodding—A method that consolidates concrete by inserting and removing a long, blunt-ended wooden or metal probe several times. (See *Vibrating.*)

Rise—A vertical distance.

Run—A horizontal distance.

S

Sand—Soil particles with a diameter between 0.08 mm and 2.00 mm.

Sash Block—A concrete masonry unit with a vertical notch cast into one end. The notched side of the block is aligned vertically at window openings and is used to secure windows in position.

Section Drawing—A drawing in a set of plans that describes the interior features of an object with visible lines.

Screeding—The act of leveling off concrete as soon as it is poured by sawing a long board back and forth on the surface of the concrete using the top of the forms and the reference stakes as control points for the top of the slab.

Screen Grid Wall—A foundation system created with insulating concrete forms (ICFs) that produces a wall with closely spaced and unique horizontal and vertical concrete members. (See *Cores, Insulating Concrete Forms.*)

She Bolts—A form tie and spreader device used with waler rods. (See *Waler.*)

Shear Strength—The ability of a material or structure to withstand parallel forces exerted from opposite directions.

Sill Plate—The lowest member of the structure's frame; the component that rests on the top of the foundation walls.

Silt Fence—A barrier designed to limit the movement of eroded materials, usually made of a geotextile erected to form a vertical obstruction to water leaving a building site or entering a storm drain. A properly installed geotextile fence extends from the base of a 6-inch-deep trench to 18 inches above grade.

Silty Soils—Soils that are predominately made up of particles between .08 mm and .005 mm. The particles are brittle when dry and very compressible when moist, creating a poor soil on which to erect a structure.

Slab Foundation—A system designed to transfer the loads created by a residence directly to the ground through a footing. The footing can be cast simultaneously with the floor or in a separate operation. (See *Post-Tensioned Slab, Raft Foundation, Reinforced Slab, Structural Slab.*)

Sleeve—A pipe or tube passing through the wall or slab forms through which pipe, wire, or conduit can be passed after the forms are stripped.

Slope—An object that has both a horizontal and a vertical component.

Slump—A simple, convenient measure of the consistency of freshly mixed concrete.

Snap Ties—Devices that hold form walls a specified distance apart. The ends are snapped or broken off after the forms are removed.

Soil Cover—A barrier to prevent the movement of water vapor in the earth from migrating into the floor of a residence. A layer of .006 mil polyethylene film commonly is used for this purpose. (See *Polyethylene Film, Vapor Barrier.*)

Spacer—See *Filler.*

Span—The distance between supporting members.

Spot Footings—An isolated feature designed to distribute a point load to the soil; typically placed under a load-bearing interior column or fireplace.

Spread Footing—The lowest part of the foundation system designed to distribute the load of the residence over a large area.

Spreader—A device used to keep two parallel forms equally spaced.

Squaring the Corners—A procedure used to verify that lines, trenches, or walls are laid out perpendicular to each other. A distance is measured 3 feet along one side and 4 feet along the other side from the same initial point. If the diagonal distance between the ends of the two lines is 5 feet the lines meet at a 90° angle. Any multiple of 3-4-5, such as 6-8-10 or 12-16-20, can be used.

Stemwall—A short vertical wall supported on a footing or other load-bearing surface.

Stepped Footings—The placement of footings at different levels. The steps should be no greater than 24 inches and each level must be connected with a vertical column.

Stirrups—A piece of reinforcing steel that has been bent, commonly into a U shape, and used to correctly space horizontal reinforcing steel in a grade beam. (See *Grade Beams*, *Rebar*.)

Story Pole—A rod or stick cut to length to provide a quick reference of the depth of an excavation.

Stretcher Block—A concrete masonry unit that does not have a flat end. These blocks cannot be used at corners or other openings.

Strongbacks—Vertical pieces of dimensional lumber used to brace forms.

Structural Fill—A noncohesive material such as sand or gravel that can be compacted easily.

Structural Slab—A cast-concrete member that integrates reinforcement steel.

Subgrade—The soil or rock below a slab.

Subsidence—The settlement of a soil mass resulting from (a) an internal cause such as natural consolidation or (b) an external cause such as the pumping of fluids, such as water and oil, from the ground. Sinkholes and "lost ground" are variations of subsidence. Sinkholes occur over abandoned mines, limestone caverns, broken water pipes, and leaking drains. One example of lost ground is the spontaneous migration of soil into an excavation.

Subsoil—A layer of dirt directly under the surface.

Sump or Sump Pit—A reservoir located in the basement floor to collect water caught in the footing drains. (See *Sump Pump*.)

Sump Pump—A pump used to move water collected in the sump pit out of the residence. (See *Sump or Sump Pit*.)

Swale—A channel whose width is several times its depth.

T

Tap Fee—A monetary charge a municipally imposes for connecting to water and sanitary sewer systems.

Taper Ties—Steel rods with one end larger in diameter than the other.

Tendons—Steel cables used in certain cast-in-place structural concrete slabs. (See *Post-Tensioned Slab*.)

Tensile Strength—An object's resistance to bending. (See *Plasticity*.)

Test Cylinders—A cylinder cast from concrete that is broken after the concrete is set to determine the compressive strength of that batch of concrete. (See *Compressive Strength*.)

Thermal Mass—An object that is used to collect and store heat gained from the radiant energy of the sun; used as a part of a solar energy system.

Tiles—A name for drainage pipes still used in some regions of the country. (See *Drainage Tiles*.)

Top Plate—The upper horizontal member of the walls.

Trench Drain—A subsurface drainage system that includes a pipe placed in an excavated trench.

Trench—An excavation that is narrower than it is deep; the open area between the foundation walls and the side of the excavation. Precautions must be taken to safeguard all workers in a trench.

Trough Block—A concrete masonry unit designed to create a horizontal void when laid in a course in a wall. The void is used to hold a reinforcing steel dowel encased in concrete.

U

Unbalanced Backfill Height—The difference between the height of the interior and exterior finished grade.

Union—A plumbing connection that is used to join two sections of pipe.

V

Vapor Barrier—A system placed under a slab to keep soil moisture from penetrating the slab. Usually composed of a bed of compacted sand or gravel at least 6 inches deep covered by a layer of either 6-mil polyethylene or 4-mil polyvinylchloride (PVC) sheets.

Vapor Retarder—A layer of material that impedes the movement of moisture through a wall or floor.

Vertical Reinforcement—Steel dowels placed vertically in cast concrete walls to increase their resistance to lateral loads.

Vibrating—A procedure used to consolidate materials. Concrete freshly placed in forms should be vibrated to eliminate trapped air and bring the concrete into intimate contact with the surface of the forms. Materials used as fill can be vibrated to diminish the likelihood that the materials will subside. (See *Rodding*.)

Vibrator—The mechanical tool used to vibrate concrete.

W

Waffle Grid Wall—A foundation system that is created with insulating concrete forms (ICFs) that produce a solid concrete wall with webs of thin, 2-inch minimum, and thick closely spaced horizontal and vertical concrete members. (See *Cores, Insulating Concrete Forms.*)

Waler—A horizontal member used to align and brace concrete forms or piles.

Wall Height—The height of a wall from the underside of the bottom plate to the topside of the top plate.

Wall Ties—Metal, fiberglass, or plastic devices used to uniformly space forms and to which temporary bracing can be secured in place.

Water Table—The level of the surface of the groundwater.

Waterproofing—The system for stopping the flow of water through a building material even when the water is under intense static or dynamic pressure.

Wearing Surface—A surface designed to take abuse.

Web—A cast-concrete wall segment, a minimum of 2 inches thick, that connects the vertical and horizontal members in a waffle grid wall system formed with certain insulating concrete form (ICF) systems. (See *Cores, Waffle Grid Wall.*)

Weep Hole—A hole placed in a wall to let moisture pass through.

Y

Yield Strength—The ability of steel to withstand a tensile load, usually measured in pounds per square inch (psi). (See *Tensile Strength.*)

Index